The Urban Book Series

Editorial Board Members

Fatemeh Farnaz Arefian, Bartlett Development Planning Unit, University College London, London, UK

Michael Batty, Centre for Advanced Spatial Analysis, University College London, London, UK

Simin Davoudi, Planning & Landscape Department GURU, Newcastle University, Newcastle, UK

Geoffrey DeVerteuil, School of Planning and Geography, Cardiff University, Cardiff, UK

Andrew Kirby, New College, Arizona State University, Phoenix, AZ, USA

Karl Kropf, Department of Planning, Headington Campus, Oxford Brookes University, Oxford, UK

Karen Lucas, Institute for Transport Studies, University of Leeds, Leeds, UK

Marco Maretto, DICATeA, Department of Civil and Environmental Engineering, University of Parma, Parma, Italy

Fabian Neuhaus, Faculty of Environmental Design, University of Calgary, Calgary, AB, Canada

Vitor Manuel Aráujo de Oliveira, Porto University, Porto, Portugal

Christopher Silver, College of Design, University of Florida, Gainesville, FL, USA

Giuseppe Strappa, Facoltà di Architettura, Sapienza University of Rome, Rome, Roma, Italy

Igor Vojnovic, Department of Geography, Michigan State University, East Lansing, MI, USA

Jeremy W. R. Whitehand, Earth & Environmental Sciences, University of Birmingham, Birmingham, UK

The Urban Book Series is a resource for urban studies and geography research worldwide. It provides a unique and innovative resource for the latest developments in the field, nurturing a comprehensive and encompassing publication venue for urban studies, urban geography, planning and regional development.

The series publishes peer-reviewed volumes related to urbanization, sustainability, urban environments, sustainable urbanism, governance, globalization, urban and sustainable development, spatial and area studies, urban management, transport systems, urban infrastructure, urban dynamics, green cities and urban landscapes. It also invites research which documents urbanization processes and urban dynamics on a national, regional and local level, welcoming case studies, as well as comparative and applied research.

The series will appeal to urbanists, geographers, planners, engineers, architects, policy makers, and to all of those interested in a wide-ranging overview of contemporary urban studies and innovations in the field. It accepts monographs, edited volumes and textbooks.

Now Indexed by Scopus!

More information about this series at http://www.springer.com/series/14773

Emanuele Giorgi

The Co-Housing Phenomenon

Environmental Alliance in Times of Changes

Emanuele Giorgi
School of Architecture, Art and Design
Tecnologico de Monterrey
Chihuahua, Chihuahua, Mexico

ISSN 2365-757X ISSN 2365-7588 (electronic)
The Urban Book Series
ISBN 978-3-030-37096-1 ISBN 978-3-030-37097-8 (eBook)
https://doi.org/10.1007/978-3-030-37097-8

© Springer Nature Switzerland AG 2020
This work is subject to copyright. All rights are reserved by the Publisher, whether the whole or part of the material is concerned, specifically the rights of translation, reprinting, reuse of illustrations, recitation, broadcasting, reproduction on microfilms or in any other physical way, and transmission or information storage and retrieval, electronic adaptation, computer software, or by similar or dissimilar methodology now known or hereafter developed.
The use of general descriptive names, registered names, trademarks, service marks, etc. in this publication does not imply, even in the absence of a specific statement, that such names are exempt from the relevant protective laws and regulations and therefore free for general use.
The publisher, the authors, and the editors are safe to assume that the advice and information in this book are believed to be true and accurate at the date of publication. Neither the publisher nor the authors or the editors give a warranty, expressed or implied, with respect to the material contained herein or for any errors or omissions that may have been made. The publisher remains neutral with regard to jurisdictional claims in published maps and institutional affiliations.

This Springer imprint is published by the registered company Springer Nature Switzerland AG
The registered company address is: Gewerbestrasse 11, 6330 Cham, Switzerland

For Sandro and Letizia.

Preface

The purpose of this book is to provide a contribution to developing critical thinking on the subject of co-housing, a fortunate term that refers to a contemporary form of vision of living and which, on the other hand, refers to a label that too often is trivially applied to residential projects, with the hope of making them more innovative and attractive. This uncertainty led me to start this research, because I believe that co-housing is an important tool in allowing society to face contemporary challenges. The book, therefore, aims to shed light on this phenomenon, defining what the essential peculiarities are for a residential project that can be defined as co-housing and presents the different characteristics that this way of life can take. The uncertainties in this field are because there can be many reasons that lead to the realization of a co-housing project, and the organizational profiles that these projects can take can also be numerous. For this reason, the work is structured as a guide to moving in this complex world of contemporary shared life, with the aim of acting as a tool to clarify the existing landscape. However, the goal is also to leave the readers free to develop a personal point of view on this topic, so that they can give their personal evaluations of the projects.

Alongside this interpretative and reading freedom, it is also necessary to express clarity with respect to a fundamental point of co-housing projects and one that is intrinsic to their essence: the ability to develop environments with a strong social potential, an indispensable requirement in the contemporary world. It is precisely in reference to this theme of strong generation of social bonds that the book affirms the importance of thinking about the role of co-housing today, and to explain this path it presents, in the initial part, a reflection on the profound crisis in which mankind stands today.

The uncontrolled development of technology, the complexity of urban life, globalization, the crisis of states and traditional communities are examples of the emergency in which contemporary humanity finds itself. It is a crisis due to the difficulty of contemporary man to establish solid, right bonds of relationships. The alarming problem is that this crisis not only refers to the social environment but also to the natural one, which has never been so threatened by a senseless and blind attitude on the part of humanity. In this situation the reweaving of social relations would allow us to create an ecosystem in which humanity can renew its existential and moral position. In this work, the environment is always seen in its interactive social and natural dimensions.

Therefore, the book presents co-housing as: (1) an innovative response to the need, which more and more people feel, to rediscover a relational environment holding down that strong individualism, typical of the contemporary society; (2) an opportunity to create a social fabric capable of facing contemporary global challenges. This perspective explains the crucial role of co-housing, which has the potential to regenerate the alliance between humanity and both the environments.

To better understand the role that co-housing can play in the contemporary world, one section of the research is the study of shared housing practices in traditions we can find in all cultures. Unfortunately, the economic and social system of the last centuries contributed to rewriting public and environmental rules, playing a part too in modifying the relational dimension concerning housing practices. For this reason, today, innovative strategies in the architectural field, in urban planning and design, can and must facilitate the development of this social renewal, regenerating ancient community housing proposals and transposing them into the contemporary world. Therefore, rewriting the alliance between humankind and the environment can pass through rediscovering socially and ecologically sustainable ways of life, which are precisely the community atmospheres, such as co-housing.

This work is therefore addressed to all those people interested in the subject of co-housing and who find themselves lost with respect to the varied characteristics that the phenomenon can assume; to those people who are interested in reflecting on its importance in the contemporary world and to the readers who want to be acquainted with the numerous commercial proposals that are offered by the co-housing brand; and to architects, urban planners and citizens interested in the reconstruction of the alliance between man and the environment through the redefinition of design strategies for the built environment.

With these aims in mind, the book is structured in three parts: the first part supports the reflection on the role that the community principle assumes in the contemporary crisis; the second part investigates the theme of shared life as an incubator of community both in historic and current forms; and the final part aims to offer insights into the topics dealt with, while also giving a hopeful vision for the future.

It should also be clear, from the very beginning, that the structure of this book is based on a research that observes phenomena from a theoretical perspective that relates the "macro" scale of global phenomena that shape the relationship between humanity and environment, and the "micro" scale of everyday life, with a particular focus on the theme of the residential trend of co-housing. In line with this point of view, the phenomena that occur on the micro scale of everyday practices affect and, in turn, are affected by phenomena on the macro scale of global trends. This theoretical framework of macro and micro scales moulds the contents and the whole book's structure (Fig. 1).

Chihuahua, Mexico   Emanuele Giorgi

Preface ix

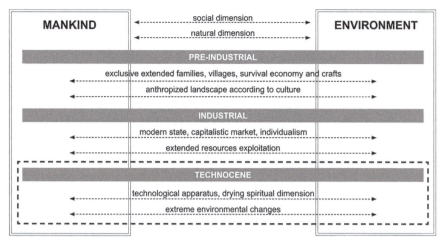

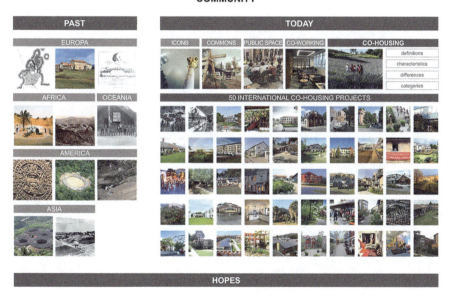

Fig. 1 Scheme of the book's structure (author Emanuele Giorgi)

Acknowledgments

This book arose from the doctoral research developed by the author at the University of Pavia (Italy), under the supervision of Prof. Zheng Shilling (Tongji University), Prof. Angelo Bugatti (Tongji University), and Prof. Fabio Casciati (University of Pavia). In the following years, the research was constantly revised and improved, collecting new material and updating information. For this activity, special gratitude also goes to all those who have provided valuable sources and comments on earlier versions of this book, particularly to Prof. Tiziano Cattaneo, Giorgio Davide Manzoni, Alessandro Giorgi, and Kenia Ivet Bravo Molinar. The Deans of Tecnologico de Monterrey, Alfredo Henry Hidalgo, Rodolfo Manuel Barragán, Jorge Gerardo Salinas, Diego Grañena, and Pablo Hernández Quiñones, supported the research activity of the last few months.

I would like to thank all the members of the co-housing communities who cooperated providing graphical material and information, without whose efforts this book could not have been written.

Contents

1 Technocene 1
  1.1 Today's Crisis: Definitions 1
    1.1.1 Environment 1
    1.1.2 Crisis 3
    1.1.3 Progress 5
  1.2 Societies and Environment 6
    1.2.1 Greece: Nature, Immutability, and Cyclicity 6
    1.2.2 The Judeo-Christian Creation 8
    1.2.3 Civitas 10
  1.3 Science and Industry 12
    1.3.1 Time, Market, and State 13
    1.3.2 Dominator Et Possessor Mundi 14
    1.3.3 Religious Technology 15
  1.4 Anthropocene 15
    1.4.1 Definition and History of the Term: A Change
        in the Perspective 15
    1.4.2 Earth Overshoot 18
    1.4.3 Responsibility Toward the Future 18
  1.5 A New Term 19
    1.5.1 Plunder of the Earth and Social Drying 19
    1.5.2 Economy and Environment 19
    1.5.3 Technique and Technological Apparatus 21
    1.5.4 What Can We Do? Ethics in the Age of Technology 22
    1.5.5 Technocene: A Definition 24
  References 24

2 Sharing as Cultural Revolution 27
  2.1 Communitarian Environment 27
    2.1.1 Sharing 28
    2.1.2 The Common Good 30

xiii

	2.1.3	Community	31
	2.1.4	Industrial Community by Adriano Olivetti	34
2.2	Creating a Human Environment		36
	2.2.1	Rationality	37
	2.2.2	Architecture and Soul	39
	2.2.3	Inhabitants Live with Soul	41
	2.2.4	The "Principle of Responsibility" in Architecture	42
2.3	Five Architectural Ways to Create a Community		44
	2.3.1	Commons	45
	2.3.2	Public Spaces	46
	2.3.3	Icons	48
	2.3.4	Co-working	50
	2.3.5	Contractual Communities	53
References			56

3 Sharing as Cultural Preexistence 59

3.1	Learning from History	59
3.2	Europe	61
	3.2.1 Polis	61
	3.2.2 Cascina	64
	3.2.3 Phalanstery	66
3.3	Asia	68
	3.3.1 Tulou	68
	3.3.2 Kibutz	71
3.4	Africa	74
	3.4.1 Matmata	74
	3.4.2 Umuzi or Kraal	77
3.5	America	79
	3.5.1 Calpulli	79
	3.5.2 Raramuri Settlements	81
	3.5.3 Shabono	84
3.6	Oceania	85
	3.6.1 Marae	85
3.7	Conclusions	88
References		89

4 Co-housing 93

4.1	We Still Need to Live Together	93
4.2	Recent Solutions	95
4.3	From Communitarian Micro-Dimension to Social Macro-Innovations	98
	4.3.1 Contemporary Territories	99
	4.3.2 Reuse	101
	4.3.3 Resilience	102
	4.3.4 Designer	104

	4.3.5	Real Estate	105
	4.3.6	The Reverse: Gated Communities	106
4.4	Common Characteristics		107
4.5	Six Features to Define "Co-housing"		109
4.6	Three Main Differences		110
4.7	Four Categories Help in Describing the Realities		111
References			114

5 Co-housing Cases 117

5.1	Studying the Cases		117
5.2	The 50 Cases		119
	5.2.1	Belterra	120
	5.2.2	Bloomington	123
	5.2.3	Cannock Mill	125
	5.2.4	Casa Tucuna	129
	5.2.5	Chiaravalle	131
	5.2.6	Coflats	135
	5.2.7	CoHousing Israel (CHI)	138
	5.2.8	Copper Lane	139
	5.2.9	CosyCoh	140
	5.2.10	Cranberry Commons	144
	5.2.11	Doyle Street	145
	5.2.12	Drivhuset	149
	5.2.13	Earthsong Eco-Neighbourhood	150
	5.2.14	Ecosol	154
	5.2.15	Emerson Commons	157
	5.2.16	Forgebank	159
	5.2.17	Frog Song	162
	5.2.18	Il Mucchio	165
	5.2.19	Itaca	168
	5.2.20	K1	169
	5.2.21	Le Torri	171
	5.2.22	LILAC Grove	174
	5.2.23	Los Portales	179
	5.2.24	Milagro Co-Housing	181
	5.2.25	Moora Moora	183
	5.2.26	Mount Camphill	188
	5.2.27	Munksøgård	190
	5.2.28	Mura San Carlo	192
	5.2.29	Nubanusit	195
	5.2.30	OWCH	198
	5.2.31	Pacific Gardens Co-housing Community	199
	5.2.32	Pioneer Valley	203
	5.2.33	Pomali	205
	5.2.34	Quattropassi	209

5.2.35 Quayside Village	211
5.2.36 Radiance Cohousing	213
5.2.37 Rancho La Salud Village	215
5.2.38 Rocky Hill	221
5.2.39 Solidaria: San Giorgio	224
5.2.40 Springhill	225
5.2.41 Stolplyckan	228
5.2.42 Sunflower	231
5.2.43 Swan's Market	234
5.2.44 Temescal Creek	236
5.2.45 Terracielo	241
5.2.46 Urban Village Bovisa	243
5.2.47 Vaubandistrict	246
5.2.48 Wandelmeent	247
5.2.49 Windsong	249
5.2.50 Wolf Willow	252
5.3 Comments on the Cases	255
References	256
6 Hopes	259
References	263
Index	265

Chapter 1
Technocene

1.1 Today's Crisis: Definitions

Although mankind has apparently become the owner of the Earth and is able to choose its destiny, the world is facing violent crises, not related just to the economic and environmental fields but also involving culture and the meaning of life. The scientific community and everyone in daily experience fully agree on the fact that today humanity is facing important changes that can be easily defined as momentous and affect both the individuals, in the intimacy of their own moral, and the society, increasingly uncertain and changing. These problems have also upset the natural and relational environment of humans, who, even if reached a point where they have unique technological capabilities, seem more and more to be the victims of a widespread unease produced by anxiety, loss of meaning, and the withering of social relations.

The successes of science, markets, and economy have provided a huge capacity to a humanity that, notwithstanding, is increasingly in a state of malaise, exacerbated by its inability to manage and to know how to use these new powers, like a modern "Sorcerer's Apprentice." High finance imposes on governments its rules, multinational corporations escape the interests of people, and technology runs without people being able to decide where to direct it.

1.1.1 Environment

In such conditions, the environment has also radically changed, both the physical one, i.e., our planet, and the social one. In the name of the imperatives of development and growth, we shamelessly exploit the planet and its resources, slowing just when we risk undermining economic profit. The imprint of man is no longer the old

---

© Springer Nature Switzerland AG 2020
E. Giorgi, *The Co-Housing Phenomenon*, The Urban Book Series,
https://doi.org/10.1007/978-3-030-37097-8_1

Sophocles' plough,[1] but rather the desert crops[2] of the Arabian Peninsula and the plantations of the Amazon forest. Also, cities, metropolises, and territories are changing their appearance, racing to become attractive city brands that are globally recognized. New buildings, the emblematic work of "archistars," draw the city skyline, which needs a new look to attract the interests of finance.

The *Oxford English Dictionary* [2] includes some useful meanings of the term "environment":

- The area surrounding a place or thing; the environs, surroundings, or physical context
- The natural world or physical surroundings in general, either as a whole or within a particular geographical area, especially as affected by human activity
- The social, political, or cultural circumstances in which a person lives, especially with respect to their effect on behavior, attitudes, etc.; (with modifying word) a particular set of such circumstances.

Following these meanings, it is easily understandable as the "environment" is the physical context surrounding man—and that is under the influence of human actions—and the context of the human relationships in which the individual is part. Therefore, environmental crisis can be easily interpreted, nowadays, as a crisis with both a physical and a social context.

As will be seen later in this chapter, every society has its own way of relating to the environment, depending on its beliefs. Within an analysis of some of the most significant societies, it is clear that the development of technology changed these relationships by taking us to a different cultural attitude. However, in a world dominated by technocrats and rationalizing paradigms, we must not consider the environment as a set of single elements that constitute the world. We must remember that the world is a system of flows, and everything has a meaning and consistency. Even thought (physical, philosophical, mathematical, etc.) has recently moved from a Newtonian, mechanistic view of nature to a network vision, made up of relationships for which the meaning and the value of an item, separate from the whole system in which it is inserted, cannot be deeply grasped.

A tragic, relatively recent event shows how the relationship between man and environment is a cultural correlation and how communitarian culture is the strongest instrument that can enable us to recover the alliance with the environment.

On December 26, 2004, Boxing Day, the Indian Ocean was the scene of one of the most devastating natural disasters in modern history that killed more than 230,000 people in 14 countries[3] [3]. A violent earthquake[4] caused a tsunami and

---

[1] As represented by Sophocles in *Antigone*: "He wearies, turning the soil year after year/By the plough using the offspring of horses."

[2] In many areas "paleowater" is being tapped to grow crops in desert environments that could otherwise never support agriculture [1].

[3] It is assumed to have been the most expensive tsunami in terms of human lives in the entire history of humanity.

[4] With a magnitude of 9.3—the third most violent recorded in the last 50 years.

1.1 Today's Crisis: Definitions

abnormal waves of up to 15 m high that hit the coastal areas of Asia. While immediately a lot of news of what happened in the more developed areas, destinations for touristic travels, was reported, nothing was known about traditional settlements in the remotest islands of the Indian Ocean. In the belief that they were even more vulnerable than the inhabitants of the "developed" coasts due to their technological backwardness, they were given up for lost, but among these natives almost no deaths were in fact counted. In an article in the Italian daily newspaper *La Stampa*, Mario Tozzi, a geologist, highlighted two elements that contributed to the safeness of the indigenous people in the most remote islands. Both concern the relationship with the environment in the two dimensions that we have just highlighted. First, there is the important component of collective historical memory (the social environment), so the population is aware that houses should not be built near the coast, and when the tide does not follow the normal trend, moving quickly into the backcountry is the best idea. Secondly, there is the preservation of the natural environment, so that the coastlines are not destroyed by wild urbanization. The tsunami caused relatively "little" damage in the Maldives,[5] which is protected by coral reefs, or in the southeast areas of India and in the territories of Tamil Nadu, where the tangled mangrove forests protected the inland areas from the devastating fury of the waves. In the local language the mangrove tree is in fact known as *the tree that defends from the wave*.

That day was a tragic lesson, of devastating proportions, about the fact that in the relationship with the environment, the modern society is totally wrong. Preserving our knowledge of the environment, transforming it in a conscious way and respecting its existence has always been, and should once again become, our way of life on the planet. Failure to do this brings tragedies. We agree with Tozzi: "It was not a 'natural disaster'" [4].

1.1.2 Crisis

Etymologically, the word "crisis"[6] designates the time to decide; it is closer to the idea of making the right choice rather than the idea of "disaster" or "catastrophe." In the contemporary condition, humankind is increasingly in situations of "crisis," in which he must make decisions because the world around him, the environment, is changing and is posing challenges that have never been addressed to him. None of the emerging entities seems to be destined to consolidate and last: everything seems to change. Without clear references, the decision-making is even more difficult, if not impossible. In past generations, every crisis situation could be solved, with reasonable hope, by actions within human capacity.

---

[5] Here there were 75 deaths compared to the tens of thousands of victims in the north-east among Indians and Sri Lankans.

[6] Crisis, from Latinized form of Greek *krisis*, "turning point in a disease," literally "judgment, result of a trial, selection"; from krinein, "to separate, decide, judge"; from PIE root ∗*krei-* "to sieve, discriminate, distinguish" [5].

This research studies and defines the general context in which modern individual moves, to propose solutions that the design can realize to help him regain his position as the goal of his actions. We must not be afraid to think of the happiness and the goodness of humankind as the goal of our activities. Society must have the strength to ensure a fulfilled life, in harmony with the environment, rich in authentic social relationships and devote both to the materiality and to the spirituality of people. Furthermore, architecture has a huge responsibility: architects have regained a leading position in the "exhibition society"; they have become popular by offering new works to cities, different from anything that already exists—works that follow the laws of marketing, of clamor. Architecture assumes an important task for the contemporary world and a responsibility for the future.[7] It must escape as soon as possible the current destruction of the territory that is perpetuated through architectural self-referencing interventions that exacerbate (and that are justified by) the fragmentation of the city. Architecture must avoid interventions that arise in the context as spaceships that are placed on the ground seamlessly with the environment, facing ecological and social issues within a limited range, and creating small clean microclimates, privileged citadels escaping a common fate. "Learning from nature," [7] the message sent by Portoghesi,[8] means conceiving the intervention as an aspect of the undivided whole, as a node, like the joining of a continuous network, capable to self-regulating at all the levels of its structure. "Learning from nature" also means no longer considering the Earth as a helpless base on which to put the numb and dumb weight of constructions.

In this vision, technology will not be the magic bullet on which to uncritically rely to solve human problems. The "smart city" (nowadays a fashionable term to indicate the city we imagine) cannot be based only on technology but rather it must come back to being the fruit of hope, of dreams, and of human culture. We need a completely new architecture, which declares itself available for a deep revision of its current directions, to point toward an alliance with the environment. The only way to rediscover this is a cultural revolution, which has the strength to bring back technology and economy to the dimension of tools, and to harden man and the environment to the more important realities to pursue. We must understand that architecture can revive the fortunes of humankind by helping to abandon the fatalistic or optimistic attitude of the technocrats, who see in globalization and in technological dominance a rosy future of progress without problems.

The relationship with the environment recognizes in architecture and ecology[9] the key role of rediscovering the alliance with a view of responsibility toward future generations. The outside world is not "no man's land" but the common good—the most important common good.

---

[7] The same is defined by Jonas in *The Imperative of Responsibility* [6].

[8] Portoghesi [7], Italian architect.

[9] Not the "ecology" that stops its attentions at the traditional positions of the environmental comforts or of the healthy home, which gives to some persons the illusion of a clean air, while consuming nonrenewable energy, thereby exacerbating the condition of the outside world.

1.1 Today's Crisis: Definitions

From this point of view, the most revolutionary aspect of the paradigm shift will be the social one, considered as an additional check of the environmental quality. After the reduction of consumption and of the ecological footprint, sustainability amounts to the ability to provide man with spaces in which he can realize his natural social vocation. In the community dimension, architecture can regain its prestige and its natural role of art at the service of man. A careful design is a strong cultural act, able to create spaces in which man can rediscover his alliance with the environment, creating and sharing social relations. Our discipline, therefore, can and must serve the good of man and his happiness.

1.1.3 Progress

The idea of progress is central to this research. In the contemporary world, "progress" now means the threat of a relentless and inescapable change that does not promise any more peace and relief but rather crisis and continued difficulties.[10] From being synonymous with hope, it has become an endless competition, and with this shift in meaning all the fears of contemporary man appear more real and they accompany him on his way without a clear direction. "Progress," therefore, has moved to the opposite end of the axis of expectations and is characterized by dystopia and fatalism.

The idea of progress concerns the direction of the civilization's development and its purpose or, in other words, its destination. The question is not whether the life of humanity constitutes a process, but rather what the direction of this process is and the end to which it tends. The issue, implicit in the idea, is the possibility of qualifying in positive terms the direction of the historical process.[11] Being a concept closely related to the destiny of man, the idea of progress has not always been present in Western societies, nor in the classic Greco-Roman culture, nor in the medieval Christian world, nor in the European Renaissance.[12] The idea has developed into a close relationship between modern science, rationalism, and the struggle for

---

[10] This is the change of meaning, analyzed by Zygmunt Bauman, in *Liquid Times: Living in an Age of Uncertainty* [8].

[11] Definition of "progress" by Pietro Rossi in the preface to the Italian edition (1964) of J. B. Bury's book *Idea of Progress* (1920) [9].

[12] Plato's theory, formulated in *Republic*, of history as the decay of humanity from an original state of perfection, and the cyclical view of the universe, received and disseminated by Stoicism, represented insurmountable obstacles to the development of the idea of progress. In fact, this idea appears only in some rare cases, for example in the Epicurean school or in Lucretius, but always in a doctrinally unfinished structure. Through the Christian principle of divine providence, the medieval conception of history as a succession of events, ordered by the divine will, is made possible. However, the concept of progress is still latent and blocked by a conception of humanity as a "damned mass." Then, in Bury's view, if men were convinced that the Greeks and Romans had reached the maximum of civilization, until their authority was considered indisputable, a theory of degeneration, which excluded any possibility of progress, held the field.

political and religious freedom. In the nineteenth century, progress was configured as the course of mankind toward a higher level of existence—from savagery to civilization. The use of reason as the directing force of human action is presented as the fundamental condition for the progress of civilization. But this implies, however, that progress is not an inevitable process, but rather a possibility that can be realized only through the work of people.

Moreover, speaking about progress implies a value judgment that is not implicit in history, but that requires the determination of evaluation criteria [9]. Therefore, qualifying an event as "progressive" or "regressive" depends on taking specific valuation parameters, which then allows the concept of progress to be used to understand history.

1.2 Societies and Environment

The relationship between humankind and environment is grounded on cultural bases and has constantly changed with the evolution of thought. The research considers myth, and then Greek and Judeo-Christian thought, emphasizing the consequences of this evolution for the city, the territory, and human life.

Myth, which allowed the first groups of people to grow and strengthen community life, describes this ancient relationship very well, but with humanism and scientific thought man is removed from his central position, grounded on a religious conception, and, at the same time, the scientific method provides him with new means, making him *Dominator et Possessor Mundi* ("sovereign and owner of the world").[13] This has been totally realized in the Anthropocene, the contemporary geological era characterized by a human imprint on the strong heart of nature.

1.2.1 Greece: Nature, Immutability, and Cyclicity

In the mythological narrative, nature, considered evil or charitable, it does not matter, has always been seen by man as a reality on which its sustenance depended, and nothing could be done to change it without asking for help from a superhuman force [10]. Fantasy and myth have allowed humans to imagine collectively. They have conferred on man, therefore, the ability for several individuals to cooperate and to do so in a non-schematic way, something that is impossible in any other species. The shared myth has enabled a scale growth that has allowed humankind to hold together empires and kingdoms, and to develop a strong sense of belonging to urban communities, from the *polis* to the *civitas*. A unifying emotional strength and a mighty tool of social control, what myth did not state was considered unimportant,

---

[13] Words ascribed to Cartesio (1596–1650).

1.2 Societies and Environment

and therefore it was not necessary to investigate anything concerning these aspects. Everything was based on a circular system, on the idea that what had been before would have to come back. This can be clearly seen in the rituals associated with these beliefs: from unexpected and unforeseen events (such as calamities) to recurring ones (changes of seasons), the ritual intervenes to make sure that everything happens again as it always has [11]. Compared with religious systems, which we will consider subsequently, it is important to highlight here a structural aspect of myth, which always refers to a bygone era from which men take the rules for their behavior in the contemporary world. Myth is nostalgia, religion is hope, two quite different anthropological categories [12].

In Greece, starting from the seventh century BC, something revolutionary happened. A social life in the *polis* allowed the Greeks to develop a systematic way of thinking, a philosophy, based on logical thinking and on research to reorganize the view of the world in a rational manner. It is a critical step according to Delort, who reports that the Greeks contributed in developing a new vision of the environment, made possible by the passage from "mythical thought to logical thinking," in turn made possible by the development of a new flexible language and by the acquisition of "rational" knowledge brought from the Middle East civilization [10]. For the Greeks, nature, considered an immutable background, was governed by a unifying principle of balance, isonomia, from which one could derive laws to found a town and for its good governance.[14] In the age of philosophy, the origin of technique, of its power over the environment, is still explained through myth. In his dialog *Protagoras*, Plato describes the myth of two Titan brothers, Epimetheus and Prometheus. Commissioned by Zeus to distribute to all the species, just after their appearance on Earth, the necessaries to survive, Epimetheus (literally *he who is not provident nor prudent*) forgets about men; to compensate for Epimetheus's failing, Prometheus (*he who is provident*) steals practical knowledge from Athena and fire from Hephaestus to give them to men. For Aeschylus,[15] the tragedian author of *Prometheus Bound*, the Prometheus feat changed men from "helpless and dumb" to "masters of their own mind." The chorus of this tragedy raises in Prometheus the question, which is the basis of our discussion, of whether the technique or the "necessity"[16] that links nature to its laws is stronger. Prometheus's answer indicates that technique is weaker than the "necessity" that governs the laws of nature.

As we will see, this can be considered true for about 2000 years, until the seventeenth century, when modern science was born. This reflection on technique, as the first form of rationality, is one of the largest machines of interpretation of myth [12]. As a supplement to this narrative, which sees the fire stolen by Prometheus as the original source of human development, Lewis Mumford proposes an observation that seems particularly appropriate for our research, recalling Orpheus, legendary poet and musician, as the first benefactor of man. According to Mumford, in fact,

---

[14] Epicurus (341–270 BC).

[15] Aeschylus (Eleusis, 525—Gela, 456 BC).

[16] Known as *anestera*, a powerful bond that governs the laws of nature.

man has become human not because he has made fire one of his instruments, but because by being able to express, through art and poetry, his symbols, communion and love, thereby enriching his life, and intensifying the moments, they had a meaning for him.

Mumford writes: "Man was perhaps an image maker and a language maker, a dreamer and an artist, even before he was a toolmaker. At all events, through most of history it was the symbol, not the tool, that pointed to his superior function. Doubtless the two gifts necessarily developed side by side, since the arts themselves … need tools … for their expressions." So Orpheus must be placed on as high a pedestal as Prometheus: the musician stands for a part of man's nature that Prometheus, notwithstanding all his love of mankind, never could bring to a full development [13].

In the Greek culture, nature is immutable; in the Judeo-Christian culture, on the other hand, the concept of technique is inscribed in the form of the domination of nature that we culturally have.

1.2.2 The Judeo-Christian Creation

If, in Greek thought, man is part of nature and deduces his laws from it to govern himself and his cities, in the Judeo-Christian culture[17] man is at the height of nature, which is at his service. Nature is not an immutable background but the creation of God, and this is the decisive matter at the root of the differences. Having been created, the Judeo-Christian nature is the product of a divine will, which first created it and then gave it to man.

Nature, then, is not the effect of natural phenomena, of chaos or chance, it is not governed by any category of "necessity" or immutable laws. The natural processes of casualness are eliminated. As we will see, this had some strong implications regarding the technical development. However, we must bear in mind that in the Judeo-Christian tradition the words "creation" and "nature" have two different nuances. Pope Bergoglio, in his encyclical *Laudato si'*, dedicated to the environmental theme, makes it clear. In the Judeo-Christian tradition, the word "creation"

---

[17] Here we are facing an issue with a global view of culture, in which the differences in some aspects can be important. The vision that interests us, however, is on a large scale to understand the trends and cultural motivations in a human approach toward nature. It remains an important difference that I want to mark in the note, or a difference between the Christian God and the Greek ones. The two concepts differ primarily because in Greek polytheism the gods are extremely irrational entities, whereas the Christian God is not a myth: "it is the epitome of rationality; it is a principle of order, not a fool who makes metamorphosis, mingles with animals and plants, quarrels, is perishable," as claimed by Galimberti [12]. Msgr. Ravasi defines culture as "the anthropological category passing through the consciousness of man towards his work and his thinking." We will refer to this definition when we use this term (words of Msgr. Ravasi during the meeting "*Il creato specchio della bellezza divina*," 1st October 2011, Parma).

1.2 Societies and Environment

does not refer just to the meaning of "nature": it is related to God's loving plan, "in which every creature has its own value and significance" [14].

Also, in the Bible there is cosmology with a unifying principle, but here the overall conception of nature is not derived from experience as for the Greeks but given by revelation. In fact, Delort and Walter observe, in their *History of Europe's Environment*, that the Holy Bible conceives the dimension of the lived reality, but this can be revealed to individuals only through revelation. "Thinking allows us to observe and to reflect, to form the experience of the relationship between nature and man. And between man and other men. It provides us with the access key to the whole environment, to its actual meaning" [10].

Unlike the Greek world, in the Bible there is no embarrassment in magnifying the embodiment of human nature in its physical relationship with the Earth. *Genesis*, the first book of the Bible, is even clearer: the creation of man does not leave freedom for interpretation regarding the matter from which man is created. "Then the Lord God formed a man ['adam] from the dust of the ground ['adamah] and breathed into his nostrils the breath of life [nishmat-hajjim], and the man ['adam] became a living being [nefesh hajjah]" (Gn 2,7).

Cardinal Gianfranco Ravasi, in analyzing this important step of the Bible, underlines a similar sound in the words *'adam* (man) and *'adamah* (ground), both of which have the same root, *'dm*, which in Hebrew indicates the ochre color of clay soil [15]. Following Cardinal Ravasi's analysis, we can see that God breathes only into Adam the "breath of life," "nishmat-hajjim," which is different from the breath of life (ruah) reserved for other creatures. These words would lead us to consider it a way to indicate that human life has a specific quality, superior to the rest of creation.

The neshamah/nishmat is never attributed to animals and it covers a number of high functions, which are often in connection with God. It is through this that man performs "spiritual acts" and he receives a special status in the order of creation [15]. According to Ravasi, this is the representation of introspection, of morality, of the ability to judge ourselves. Therefore, humankind is appointed "lieutenant" of God in creation and then a "royal" function to be performed is assigned to him.

Delort and Walter propose this domain concept, observing that this must be moved by the qualities of nishmat-hajjim that God gave to man. In fact, even if the mission of humankind is to dominate nature, this still remains the perfect instance of God's wisdom and people must dominate it with intellect [10].

This means that in order to administer creation, man must use his conscience. This idea of a link among religion, morals and the protection of creation and landscape can be found in the etymological analysis of the Hebrew words *avad* (*cultivate* and *adoration, prayer*) and *shamar* (*safeguard* and *compliance with the law*).[18] Working on the Earth therefore means having a covenant with God, a deep and religious bond that has been given by Him, but original sin and expulsion from the

---

[18] Words of Msgr. Ravasi during the meeting "*Il creato specchio della bellezza divina*," 1st October 2011, Parma.

earthly Paradise make the work of man uncertain, problematic, because man remains free to determine and act according to his will, but nature rebels against the abuse by man.

Even in the Muslim religion, environmental protection depends on the morality of human actions. The Muslim parable about creation tells how man was placed in a large fertile garden but was warned that, for every sinful act against himself, another man or against the Earth, God would drop a grain of sand. Thinking that a grain was nothing compared to the immensity of the garden, men went on committing sins without problems, so deserts advanced incessantly.[19]

The theme of the relationship between humankind and the environment and, in particular, the issue of the problems that human behavior can cause to the environment is central in the Bible and it is, especially nowadays, given close consideration as we can read in the encyclical *Laudato si'*. This relationship between humankind and environment has got a second aspect that is to be carefully considered: technique, registered in the Western thought as a form of domination of nature.

1.2.3 Civitas

We try now to present summarily what connotation and significance the urban structures have taken in the aftermath of the classical age, a "world of cities," which declined between the third and fourth centuries after Christ. In many cases it was precisely from sites of Roman cities that the medieval ones developed, considering the classic monuments not only as ancient mysterious ruins but sources of building material first and foremost. As observed by Le Goff in Europe, vocation and functions change: unlike the previous cities, the medieval ones were not just military, political, and administrative centers, but also hubs for economic and cultural blooming [16].

In addition, the relationship that the city has with the countryside changes. It was the genesis of Western life, a real revolution, both in terms of agricultural production and in a new urban model, that passed from the ancient Roman opposition between *urbs* and *rus* to a new relation between city and country [17], even if the difference between *municipes* and *incolae* persisted in the juridical system.

In the Greek world, the space of the city was a well-defined area, separated and intended for the protection of people. The city represented the entire and only area of responsibility in human nature, not being subject to anything but the category of "necessity" [6]. With the arrival of the medieval European communes and with the Christian conception of nature that dominated them, this strong distinction between the city and the rest of the natural world vanished. However, whatever the culture, the need for protection does not disappear. So the walls take on a significant

---

[19] Characterized by a strong transcendental vision, "*you, man, you are a puddle, sometimes you can reflect the sun, but the sun is far else.*"

1.2 Societies and Environment

importance for the medieval cities too, but they are not just a technical and military structure. They take on very clear economic, social, and symbolic meanings, because by defining the outside and the inside, they represent the linking point between these two worlds and define the physical place where the separation and the meeting of these two worlds is realized [8]. The walls in the fresco by Ambrogio Lorenzetti, *The Allegory of Good and Bad Government*, in Siena, embody this complex of functions well.[20]

In this famous fresco, in one part we have a highly valued and determined space, enclosed by walls, made up of houses and a market; in the other part the fields and nature represent a subordinate space that works for the city and provides resources and products (and landscapes). This painting represents very well the exchange between town and country in the medieval communes. Moreover, according to Le Goff, if every demographic agglomeration already involves, usually, the awareness of belonging to a community,[21] this is even more true when we speak about the city, where the permanent visual presence of a boundary wall within which to live creates and consolidates the feeling of belonging to the same community [18]. Nevertheless, the medieval city was able to add value and new meanings to the countryside, to the surrounding natural environment. Nature became esthetic landscape (esthetically represented).

The Landscape of Francesco Petrarca

The climb to Mount Ventoux in Provence by the Italian humanist Francesco Petrarca, on April 26, 1335, represents the first modern attention to nature as landscape—a modern concept of landscape as a show of nature taken up by our feelings. The climb is narrated by the poet in a letter addressed to his Augustinian friend Dionigi di Borgo San Sepolcro.[22] To understand the importance of this event we must emphasize that Petrarca climbed the French mountain without any practical interest, but was driven only by the desire to freely contemplate the whole of nature and God, watching from the top of the mountain the nature below. This description shows that the idea of landscape is unknown to shepherds and farmers who live in nature. Territory does not become a landscape only when people observe nature in experiencing the pleasure of contemplation, but it assumes a value that lasts over time. For this reason, Burckhardt said that Petrarca and his brother, who accompanied him during the climb, were the first, among modern people, to perceive and enjoy the landscape from an esthetic point of view [19]. Confronted with an unimaginable view, Petrarca perceives a nature that becomes pure contemplation; the poet becomes estranged and reaches a feeling of wholeness with himself, with God and with the landscape. Nevertheless, landscape is born from detachment: the result of a

---

[20] Ambrogio Lorenzetti, *The Allegory of Good and Bad Government* (Public Palace of Siena, Italy, 1338–1339). Presented by Le Goff as a meaningful example in *Costruzione e distruzione della città murata* [18].

[21] Le Goff refers also to the *Signorie*, villages or monasteries.

[22] Dionigi had given to Petrarca a copy of the *Confessions* of Saint Augustine. Petrarca refers to having read a passage of the work of the saint, once he had reached the top of a mountain.

separation between man and nature. Between ancient times and modernity, then, there is a separation given by a modern sentimental interest in nature, which Schiller[23] noted with surprise was missing among the ancients. Moreover, this esthetic feel of the landscape is far from a philosophical theory of the cosmos in which man and nature are part of one entirety. It should be noted that in no historical age is the relationship with nature dictated by a purely aggressive attitude or by an exclusively contemplative one. These are two approaches that coexist depending on the society and on the culture of reference, so that these relationships evolve continuously on the basis of the cultural evolutions of the society. At this point, man can approach nature with two different gazes: that of the poet and that of the scientist. The contemplation of nature has played an important role in all cultures, beginning with the Greek one, which, however, does not go beyond a "deep feeling for "nature" [20].

1.3 Science and Industry

From the end of the sixteenth century a new European society developed, focused on inventions and discoveries that responded to the needs of a changing society that had increasingly more problems to be solved and at the same time had more and more tools to solve them. What is perceived is an important passage from an ancient world where work and thought were distinguished to a new world where industrial work and scientific thinking began to be combined in a fruitful alliance. The union between labor and research spirit advanced the age of autonomy and sovereignty of reason [21].

It is useful to point out that this change in thinking has unique properties that cannot be found in any other previous tradition of knowledge. These peculiarities have enabled the success of this form of thinking that led to the contemporary relationship between man and environment [22].

- Scientific thought admits ignorance in the sense that the knowledge that we possess does not cover the totality of things that can be known. Moreover, it is not necessary to say that what we know is necessarily right: progress is often reached by the successful passing of convictions through discoveries that contradict them, and that will be contradicted by subsequent progresses. This feature is the big difference with religions that, based on dogma, cannot be contradicted, or questioned.
- The collection of information by observation and by processing it through mathematical means plays a central role in the scientific discourse.
- The action of modern science does not end in the definition of new theories but goes as far as the development of new technologies, delayed effects of scientific applications.

---

[23] Johann Christoph Friedrich von Schiller (1759–1805), German poet.

1.3.1 Time, Market, and State

If what is commonly called a "scientific revolution" can be considered the expected result of the development of a scientific method, it is also true that these results have found immediate application (and have been also incentivized) by the needs of an era.

This great process that marks the birth of the industry represents a new way of converting energy. A highly symbolic invention of understanding and exploiting energy is James Watt's[24] steam engine, the first model of a condensing steam engine (1763), which represents today's "energy-consuming" world. The revolution started by this engine led to the modern dependence of Europe on energy [10].

It should also be noted, however, that the Industrial Revolution brought a second agricultural revolution because the methods of industrial production became the strong points of agriculture [22]. The application of the scientific method to the world of production even had a direct impact on the social order, both in terms of thought and as regards the perception of time and the social structures.

We have already made some observations about the importance of shared myths in structuring societies and in keeping millions of people connected. We also saw that the essence of myths is diametrically opposed to that of science: the former does not allow changes, while science is constantly evolving and, with scientific evidence, it crumbles many of these myths. If the myths are in crisis, one problem is to figure out how to keep communities, countries, and states together. In modern times, as Harari underlines in his *Sapiens: A Brief History of Humankind*, two approaches to solving this problem have been developed: either a scientific theory is considered absolute and final, as the totalitarian regimes did, or science is abandoned to embrace an absolutely unscientific truth, which is characteristic of liberal humanism [22].

Another consequence that shocked the society was the change in the perception of time. With the industrial development, "industry time" was created. It can be gained or lost, wasted or organized more efficiently. It is beyond dispute that industrialization led to a lifestyle governed by its own rules and by its prohibitions. According to Delort and Walter, the functional and geographical separation of work, as we are experiencing nowadays, and the divorce of private and professional life are all constrictions against the natural state [10].

Also, urbanization, the role of peasantry and the industrial proletariat, poverty, democratization and the "youth culture" have undergone significant changes, but, above all, the collapse of the traditional family and the local community took place, replaced by the state and the market. In fact, man always lived in restricted communities, often parental, where family and community ensured the satisfaction of needs. The technology destroyed these structures. The Industrial Revolution gave the market immense and new powers, provided the state with new means of

---

[24] James Watt (1736–1819), Scottish engineer and inventor.

14 1 Technocene

communication and transportation, and placed at the government's disposal an army of clerks, teachers, policemen, and social workers [22].

1.3.2 Dominator Et Possessor Mundi

A willingness to control and dominate nature was already present before the scientific revolution, but only in the seventeenth century, and it will have the strength and the technological tools to impose a really significant change in the relations between humanity and environment.

The power that men possess is limited to the sciences, while nature cannot be subjugated at all. This is the message that Francis Bacon launched in aphorism 129 of the *Novum Organum*. He reflects on the ambitions of science at the dawn of the scientific revolution, but he also insists on submission to the laws of nature, albeit without any reference to any form of respect for it.

The scientific discovery of the laws of nature becomes a way to know the law of God deeply, and the dominion of man over nature begins to be established.

From the seventeenth century the hypotheses of the scientific community made on the operating principles of natural events are verified through experiments whose subject is nature itself. If the experiment confirms the hypotheses made by man, these become laws of nature. There is the essence of humanism: beyond the distinction between science and the humanities, we can say that science is the essence of humanism.[25] Also, Hans Jonas is deeply convinced of the radical change that has led humanity from the ancient static cultures to a dynamism that makes people masters of the environment such as to allow each kind of consumption: with the birth of modern science and the consequent dynamism, a radical transformation of the relation between humanity and environment occurred. Humankind passed from being dominated by nature to being the dominator of nature [23]. But a natural disaster of catastrophic proportions led to greater caution about the domination of nature.

1755 Lisbon Earthquake

On November 1, 1755 a natural event upset the history of the European continent, changing the fate of one of the most prosperous countries of that time. The coastal areas of south-western Europe were hit by a devastating earthquake and a subsequent tsunami due to a tectonic movement whose epicenter was near the Portuguese coast. In particular, the most affected city was Lisbon, the capital of the Portuguese kingdom, which put the number of victims at between 60,000 and 90,000 out of a population of about 275,000 inhabitants. It is believed that collapses, fires and tsunami waves destroyed more than half of the city. In direct terms, this catastrophic event marked a heavy blow to the Portuguese society, which had to abandon any ambition of overseas colonialism. A similar event had no consequences in the

---

[25] Consideration of Umberto Galimberti, expressed during the conference "*Educare l'anima ai tempi della tecnica*", May 2010, Muro Leccese, La Bussola theatre.

geopolitical world, but strongly shocked the conscience of an entire continent and became a stimulus for philosophical and religious reflections.

A Catholic country, involved in the evangelization of new lands, was struck in the heart on All Saints Day by a natural event. The disaster offered Voltaire, the great Enlightenment writer, a reason to criticize Leibniz's idea of the existence of "the best of all possible worlds" through the misadventures of Candide, the protagonist of the eponymous novel. After the Lisbon earthquake, the arbitrariness of man's destiny on Earth is clear to Voltaire.

1.3.3 Religious Technology

In this continuous development, scientific thought is increasingly intertwined with the religious sphere and with the interpretation of the environment. It becomes a means to explain the phenomena and dominate nature. This scenario is developed until the point of upsetting in an irreversible way our environment and human cultures.

Inaugurating the scientific method, Bacon holds that man through science (and technology) will retrieve the preternatural virtue of Adam, before the original sin, reducing the negative consequences associated with the punishment: the technique will allow the hardness of work and the pain to be reduced. It should be noted, anyway, that science deals with the explanation of natural phenomena and with the attempt to change them, which is very different from the role of religion, which is called, instead, to provide an explanation of sense.

Finally, an interesting observation on the perception of time. Just as religion thinks that the past represents evil, the original sin, the present is redemption and the future salvation; science is based on the same scan time, made by a past of ignorance and assumptions to be overcome, a present of research and a future of progress. We will try later to investigate in greater depth the relationship between scientific development, politics and economics. Over the past five centuries, men thought they could increase their power through scientific research, nourishing an almost religious faith in technology and science to the detriment of revealed truths [22].

1.4 Anthropocene

1.4.1 Definition and History of the Term: A Change in the Perspective

As the scientific outlook changed the perception of the world, so the technological action changed nature. Starting from the nineteenth century, the ambition to subjugate nature, seen as a set of resources available, reached its peak. Western civilization became characterized precisely by this acting directly on the environment and on its resources with violence, injustice, and inequality [23]. In the second half of the eighteenth century, however, vegetarian movements and criticism of vivisection developed particularly in the Anglo-Saxon countries[26] and the nineteenth century began to be sensitive to animal suffering [23]. In England, hesitancy first appeared between two attitudes: the choice of culture and the choice of nature, a "human dilemma."[27] On the one hand, nature was still a source of youth and happiness, and on the other hand a contemporary hard and repressive world. The reflections of the society of the nineteenth century were developed around these two poles and progressive utopias were confident in a future bearer of solutions to free humanity from the oppression of technology and modern production.

So in the late nineteenth and early twentieth century many movements had as their goal a return to nature. Many of them were inspired by Ernst Haeckel, a German physician and naturalist, who coined the word "ecology" in 1866.[28] According to him, nature and man are only one entity, benefiting from the same moral and natural conditions. Haeckel claims that in nature, we can find the truth and rules of life. From these first movements of a return to nature, many other realities developed, undermining the idea of technological progress indifferent to ecological implications and enhancing the recognition of a natural heritage and the development of an environmental esthetic [17]. Anyway, until the end of the twentieth century the respect for the environmental balances had never been powerful enough to influence common ethics, something that happened when Westerners understood the fragility of the environment, making it not only a philosophical and scientific problem, but also a real social one. New forms of sustainable development had to be found through a policy that brought these issues to the core of its own actions to ensure the survival of the society. From the eighteenth century signals on the heaviness of human actions on the environment were launched. We propose a few of them. Louis Ramond de Carbonnières, in the late eighteenth century, described the role that men had in structuring landscapes and highlighted the destructive power of their action. "A century of human activity encumbers on the Earth more than 20 centuries of natural activity" [10, 24]. Meanwhile Jean Brunhes, a French geographer, in 1910, in writing "La Géographie Humaine," argued that "[d]evastation is a distinctive peculiarity of civilized people" [10, 25].

In 1935, Arthur G. Tansley took the credit for having coined the term "ecosystem," which means "not just the combination of organisms, but also all the physical factors that represent what we call 'environment'".[29] In this perspective, we had the perception of the global environment in which man is an infinitesimal part, inextricably

---

[26] Probably due to a greater attention to the landscape that developed in these countries (also due to a painting that no longer addresses religious subjects but is more focused on nature).

[27] Sir Keith Vivian Thomas (b. 1933), British historian, whose concept of "human dilemma" is explained in [23].

[28] Ernst Haeckel (1834–1919), German biologist and philosopher.

[29] Reported by Delort as quoted in Edward J. Kormondy, *Readings in Ecology*, (Prentice-Hall, Englewood Cliffs, 1965) [10].

1.4 Anthropocene

linked to other living beings and to inorganic components such as land, air, and water. All this comprises an ecosystem, because all the components are dynamically supportive in their relationships and they tend toward a certain stability. Therefore, human interventions are a critical factor in the evolution of the biosphere, albeit unfortunately with dangerous consequences such as acid rain, desertification and so on, while man himself inevitably remains dominated by the laws of the environment.

From numerous studies carried out after the end of the nineteenth century,[30] it was possible to define precisely the alternation of geological eras and explain the interrelations with changes in the climate of the planet, in turn closely linked to the chemical composition of the atmosphere. The alternation of ice ages and warm interglacial periods has been the rule for the last 740,000 years. Throughout this period, in the hot stages the concentration of greenhouse gases increased considerably. In particular, the so-called "trace gas" changed: the concentration of carbon dioxide increased in the order of about one and a half times, and that of methane gas of about two times. All this until two centuries ago, when there was a further growth of these gases, which reached levels not seen in the past 15 million years.

The theory of the Nobel laureate Paul Crutzen[31] is precisely that these changes in the parameters of trace gases will mark the entry into a new geological era that he intends to call the "Anthropocene" (from the Greek *anthropos*, man). This new era, unlike the previous ones, is characterized by the strong, crucial impact of man on the environment: an epoch when the planet and its elements have been modified by human action to a greater extent than the environment itself can do. We move more matter than atmospheric agents; we alter the cycles of water, nitrogen, and carbon; we create new chemical elements. The Anthropocene is marked by the action of a single species that suddenly found itself in a position to determine the balance of Earth and climate. In addition, this situation of power influencing the environment can only grow, moved by the forces that govern human action. It is impossible for us to go back, but nothing prevents man from being able to learn how to modify the effects of his actions.

Crutzen's definition is sharp, reminding us that humanity is at the core of this new era, the Anthropocene. Closing his *Welcome to the Anthropocene*, Crutzen describes this new geological era as the only one in which nature does not represent an exterior power that "rules over human destiny," but, on the contrary, in which humankind determines its equilibriums. For this reason, people are all called to act, with wisdom and responsibility [27].

---

[30] From 1840 with the researches of Christian Friedrich Schönbein on the ozone, or in the 1920s with the studies of Sydney Chapman and Gordon Dobson, up to the important probing at the permanent Antarctic base of Vostok in 1999.

[31] In 1995, Paul J. Crutzen shared a third of the Nobel Prize in Chemistry with Mario J. Molina and F. Sherwood Rowland for their *work in atmospheric chemistry, particularly concerning the formation and decomposition of ozone* [26].

1.4.2 Earth Overshoot

In 2018, it happened on 29th July: the day by which (ideally) the world's population had consumed all the resources produced (estimated) from the Earth in the same year. The ecological footprint of humankind is heavier than the biological capacity of the planet and, in seven months, we burned all the natural resources produced in 2018. Since July 30, 2018, humanity has begun to exploit the "2019" reserves, thereby compromising land, forests, seas and fauna. This means that to satisfy our needs we are using resources as if we would have 1.6 Earths and we are consuming, therefore, the "environmental capital" of the planet, making it more and more difficult for future generations to meet their needs. This is what the Global Footprint Network says [28].

The last year in which man consumed in 12 months as many resources as the planet could offer him was 1970. Of course, excessive consumption has huge repercussions for the environment. These include deforestation, advancing desert areas, a scarcity of fresh water, pollution, and a loss of biodiversity. It is a matter of land, water, air, and life.

Twenty years ago, Edward O. Wilson and Niles Eldredge wrote [27] that with the rhythms then in progress, the extinction of species because of human activities would be inevitable. Now, the biosphere is experiencing a great mass extinction, a catastrophe on a global scale.

In 2018 "Living Planet" report by the WWF [29], the Living Planet Index showed a staggering overall decline of 60% in species population sizes between 1970 and 2014.[32] This means that the number of wild vertebrates (mammals, reptiles, amphibians, and fish) in the world, in the last 40 years, has reduced by 60%. The greatest decline was recorded in tropical areas, especially in Latin America with a decrease of 89%, due to habitat destruction and to uncontrolled exploitation [29].

1.4.3 Responsibility Toward the Future

We have seen how the power of humankind has increased to the point where it has modified directly the life of our planet on a global scale: heating, gas and aerosol emissions, the loss of biodiversity and the destruction of natural habitats. Many of our activities generate heavy and often uncontrolled impacts on the planet. Their amassing does nothing but make unpredictable the consequences that, presumably, could be catastrophic. Anthropic activities have created a huge hole in the ozone, but once the problem had been identified, man managed to stop it from growing. This is the real strength of man in the Anthropocene. Man has already proved he is

---

[32] The data are based on the trend of 16,704 populations of 4005 vertebrate species across the globe [29].

capable of ensuring the preservation of his species (and other ones) through the use of technological tools and strategies that make it possible to find a new balance with the Earth. The solution to the problem is extremely difficult, but the first thing is that we should identify, on a global level, the balancing of the ecosystem as a priority in terms of human action. This must be the goal, and certainly not economics, production or material "progress." It means, as we will see later, having the strength to change the paradigm. And there are some positive signals: from the development of renewable energy sources to the practice of recycling to a more widespread environmental sensitivity.

1.5 A New Term

1.5.1 Plunder of the Earth and Social Drying

While the world we are living in is based on a continuous consumption and material development, the central role undertaken by economy and technology led to the creation of an "apparatus." The forces that have developed as a result of the scientific revolution should have made man free from every constraint, putting him above nature and allowing him to decide freely of his own accord. Technology and production developments, seen as liberating progresses, should have enabled the society to produce more, work less and be freer. Instead, these same forces, these scientific developments, seem to have brought the society into a situation in which the human dimension is lost, and man is again deprived of his freedom.

Certainly, it should be borne in mind that too often, nowadays, the consequences of human actions are no longer under the direct control of mankind. Even an asset as important as food, which is the essence of life and a strong cultural basis of communities, has now become something more and more expressionless. It is widely believed that society is following guidelines that are only partly dictated by the real will of man, and that are increasingly dependent on fulfilling the needs of economic and technological apparatus: if before the decisions could be addressed by communities and politics, today this is no longer the case.

1.5.2 Economy and Environment

That economic development depends on production has now been accepted in our society as an absolute dogma—a dogma of an economic paradigm that, having been developed since the seventeenth century, led the economy to have an unbreakable bond with material production and the environmental footprint. This connection has been the subject of studies since the second half of the twentieth century. Scholars such as Nicholas Georgescu-Roegen, Kenneth Boulding, and Herman Daly underlined the links between economics and environmental matters [30]. The stance of ecological economics is rigorous. It sees the proposals of neoclassical economics just as limited attempts, insufficient to really solve the problem, and it appeals to a radical upturning of our idea of economic process that should abandon the size of neoclassical economics and its imperatives of growth. The need for this kind of continued growth (at least for the rich countries) is challenged by the increasingly clear awareness that economic disparities are increasing and that the possibility of exploiting the environment is limited [31].

On a global scale, the economic growth with its environmental impact is weighing on the planet more heavily than it can withstand and in many countries is producing more damages than advantages. The problem with our society is that we are led to think that the aim of our actions should be linked to the economic field. On the one hand, there are new products and services, and on the other, there are our demands as consumers, driven by complicated social logics, for more and more. So we cannot but think that growth is the solution to all the ills of society; if there is growth, there is well-being, because "[u]nder existing macroeconomic arrangements, growth is the only real answer to unemployment—society is hooked on growth."[33]

In regard to the domestic economies of single countries, however, the problem relates to inequalities, as wealth is increasingly concentrated in the hands of a small number of super-rich who are able to control huge economic resources, and this inequality has worsened in recent decades [33, 34].

Ultimately, we are locked up in the "iron cage" that Max Weber describes to indicate the limits to individual freedoms in the capitalist and communist systems [32]. In modern societies there is a paradox. In fact, humanity has achieved a high level of material and technical progress but still suffers from anxiety; we are led to depression, worried about how others see us, unsure of our friendships, driven to consume time and without a life of community worthy of this name. The loss of a relaxed sociability is replaced by material excesses of food, shopping, alcohol, etc.[34] [36].

---

[33] Douglas Booth (2004), quoted in [32].

[34] The data suggested in his testimony by John Perkins, author of *Confessions of an Economic Hit Man*, are interesting. In Ecuador, during the "oil boom," the poverty level rose from 50 to 70%; the under- or unemployment level from 15 to 70%; the public debt from $240 million to $16 billion; the national resources allocated to the poorest segments of the population dropped from 20% to 6% [35].

1.5.3 Technique and Technological Apparatus

In the history of humanity this situation is an event that is unique and highly significant in its dangerousness. From being a tool to give a sense of order and security to the humankind surrounded by an unpredictable nature,[35][36] technique converted to a technology which came to threaten the fall in a re-barbarism, a final defeat in our relationship with the environment, if not even the very end of our story. Also, according to Hans Jonas, this may mean the failure of a superior culture that falls into a new "primitivization" caused by the irresponsible habit of wasting [23].

In fact, the technological civilization possesses a tendency to be uncontrollable; the technique has gone beyond its original instrumental meaning and now technology controls the living conditions of men and their goals.

Gunther Anders is a philosopher who studied the concept of technology in depth, examining the first three industrial revolutions from a historical perspective. The first revolution was characterized by an iteration process using machines that produce machines using a machinery principle. It was a chain that ended with a machine that produces products. Meanwhile the second, triggered in the nineteenth century, produced needs. Finally, the third revolution brought drastic changes, new imperatives that tended to stifle the moral postulates of the social and individual ethics of the past. According to Anders, what technology and production could achieve became mandatory and products had to be used according to the design purposes without any restriction [37]. So, from a physical discovery perspective, nuclear power became a weapon that threatened the very survival of humanity and it is now the symbol of the third industrial revolution, an era in which we manage the production of our own destruction. Moreover, as several philosophers (among them, Emanuele Severino) think, it is not a symbol of the physical value of the discovery, but of its possible effect, as its metaphysical meaning and the nuclear threat are seen as the anticipation of the planetary government of technology.

Every human characteristic that goes beyond the needs of the technology is suppressed and forgotten, leading to a state of drying. There is no doubt that humankind, after undergoing revolutionary changes in his being and finding dried his spiritual features, which do not align to the scientific apparatus, is forced to live in a state of anxiety, without points of reference and purpose for his actions.

There is another aspect that creates a sense of anguish and anxiety in us as modern people. Pain and death are dramatic and unavoidable occurrences in life, through which man can obtain salvation; they were inscribed in scenarios of religious salvation that have now collapsed and been replaced, in part, by science. We still

---

[35] While the symbolism and myth endowed Humankind with a sense of superhuman power and creativity, the scientific and technical advances, gave us a further insight into the need of conforming to nature and accepting the conditions of the environment [13].

[36] Can also be seen noted as the same term of technique, "techne," results from "Hexis Nou," that is "be master of his own mind," thus freeing it from the need of having to depend on divine explanations.

look for salvation, but not being able to find comfort in a religious background, we look for it in science, which moves proposing biological and pharmacological solutions, which cannot, in any case, eliminate the anxiety of the infinite desire for happiness.

This sense of anxiety is described by Mumford with the illuminating figure of a man who has become a stranger in a mechanized world, or, even worse, has become a "displaced person" [13]. In fact, if on the one hand, Mumford says, technical progress has given man a highly organized, orderly, and predictable environment, on the other hand something is missing in his spiritual life, "something essential to this organic balance and development," This "something" is precisely *the human person* [13].

1.5.4 What Can We Do? Ethics in the Age of Technology

At this point some questions arise: what happened to the technic and to humankind was a mistake? The direction taken by technology should be left? Can we put ourselves back in a situation of power from which we can dominate the technological development and, once again, become masters of our own destiny? Drawing a parallel with *The Sorcerer's Apprentice* by Goethe, Jonas wonders if we can take up the situation again and his pessimistic answer is that if wisdom and political judgment will not succeed in doing that, then perhaps fear will succeed [23].

Mumford gives an idea of hope instead: "Much that is now in the realm of automatism and mass production will come back under directly personal control, not by abandoning machines, but by using them to better purpose, not by quantifying but by qualifying their further use" [13]. Mumford was profoundly convinced about the power of humankind to go beyond the scientific curiosity, beyond the building of machines and to carry out regular jobs. Man, for Mumford, will be able to make new decisions and make choices that, if made before a disaster, will lead to a renewal of life.

The drying of man, the loss of happiness, and of the sense of life have an important link with the cognitive and ethical components of individuals. They are the values that give us the benchmarks that help us to understand whether our actions are correct or not and what situation we are in: if we are the slaves of a dictator or free citizens. Moreover, these same values are those that allow us to endure the pains of life; their absence makes life itself an anguish. If it is true, as we think, what Harari says, that is, that the scientific point of view removes from human life any form of sense, we understand how dangerous it is for man to adopt a purely scientific approach to life and reality [38].

We have already seen how the technological theme is dear to Pope Francis, who highlights how the growing of technical capabilities is not yet followed by an equal increase in the ability to use that power. The Pope writes, "there is a tendency to believe that every increase in power means 'an increase of progress itself', an advance in 'security, usefulness, welfare and vigour; ... an assimilation of new

1.5 A New Term

values into the stream of culture', as if reality, goodness and truth automatically flow from technological and economic power as such. The fact is that 'contemporary man has not been trained to use power well', because our immense technological development has not been accompanied by a development in human responsibility, values and conscience"[37] [14].

We have not yet been able to define an ethic at the level of the scientific happening, so scientists follow "a scientific ethics" according to which you do what you can do. If in the traditional societies everybody were satisfied with repeating what the predecessors did, our society would constantly seek new goals: it has become a dynamic society. So we can start from the consideration that the new status of technology and the central role that it has assumed in human activities require it to acquire an ethical dimension. Jonas is sure about the fact that if making becomes an essential action, then morality must, as public policy, enter the dimension of "making" [6]. What gave us the means to achieve the technical–scientific level of today has weakened at the same time the fundamental principles on which the rules of behavior were based. It is what Jonas calls an *ethical vacuum* [6].

In 1985, Jonas argued that there was no need for a new ethic, but certainly there would be a need to rethink the ethical duties. The technological and scientific progress cannot proceed alone without progress in terms of intellect and morality. In this situation knowledge becomes a duty more than ever necessary given the range of our actions. Now, with man being provided with a predictive capacity lower than potential consequences of his actions, knowledge assumes an ethical significance [39].

At this point, an important indication is offered by Agnes Heller: a person's rectitude can be seriously compromised if this person limits their attention exclusively to subjects or objects included in their range. In fact, moral interest goes far beyond the individual or group horizon. It is not only a responsibility toward the "remote" in terms of space, but also a responsibility to those who are "remote" from the point of view of time. Nature, Jonas reminds us, does not give us an innate duty toward future generations (nature does not do it and even the earlier ethics did not do it). But, he adds, we are responsible for the *idea* of man and not for future of people [6]. In *Laudato si'* Pope Bergoglio also introduces the concept of responsibility to future generations, believing that the solidarity between generations is a fundamental principle of justice, because the world that we receive from our fathers also belongs to those who will follow us [14]. To conclude this excursus on moral and ethics we recall the relationship between morality and society specified by Heller. "There is no social life without ethics; every society and every movement have their own ethical precepts" [40]. So a gap in the field of ethics will also involve considerable social problems. The rules of the traditional ethics left little space for individual decisions, based on inherited ethical rules, so that in the face of contradictory regulations, individuals could decide on their priorities [40]. The agricultural world

---

[37] Pope Francis in this passage quotes the Italian theologian Romano Guardini (1885–1968) in Guardini, R. (1998). *The End of the Modern World*. Wilmington: ISI books, p. 82.

has a community and homogeneous nature, but it is imbued with a very rigid morality. Today, daily life is characterized by a relative emptiness due in part to the fact that the sense of community is lacking, and in part to the fact that the development of individuality is locked. Moreover, the combination of community and autonomous individuality is the unfulfilled need that we must satisfy [40].

1.5.5 Technocene: A Definition

If today man is nothing but "co-historic" with technology, that means that just technology and technological apparatus draw the course of history, and it should be more appropriate to call the new geological era "Technocene" rather than "Anthropocene." The US paleo-climatologist William Ruddiman[38] introduced the idea of the "Early Anthropocene," a geological era that began with the first farming and breeding activities so many thousands of years ago, a time in which man could still be described as the author of history [41]. However, now the scientific community embraces the theory of Crutzen, who places the beginning of this new era in the mid-eighteenth century, when the industrial and scientific development entrusted the domain of humankind to technology [42].

The concept of Anthropocene was unquestionably a revolutionary and innovative idea, because it highlights the awareness, at least in the scientific community, of the explosive force of the changes taking place on the Earth. The idea of having entered a new geological era, determined by human hands, has the effect of stirring individual consciences and collective awareness. However, the term "Anthropocene" is deceptive because it still indicates that man is the driver of these changes, while with the term "Technocene" we underline how the transition to management driven by technology represents this passage to a new geological epoch.

References

1. Welcome to the Anthropocene. Explore places (2012). http://www.anthropocene.info. Accessed 29 August 2019
2. The Oxford English Dictionary. www.oed.com. Accessed 7 February 2019
3. D. Hurst, Boxing day tsunami: a survivor's story. BBC News (2014). https://www.bbc.com/news/uk-30537152. Accessed 29 August 2019
4. M. Tozzi, Quello tsunami ci ha ricordato che la natura è ancora sovrana. La Stampa (2014), p. 11
5. Online Etymology Dictionary, Crisis (2019). https://www.etymonline.com/word/crisis. Accessed 12 August 2019
6. H. Jonas, *The Imperative of Responsibility. In Search of an Ethics for the Technological Age* (Chicago University Press, Chicago, 1985)

---

[38] William F. Ruddiman, paleo-climatologist and Professor Emeritus at the University of Virginia.

References

7. P. Portoghesi, Imparare dalla natura/learning from nature. Domus **818**, 3–16 (1999)
8. Z. Bauman, *Liquid Times: Living in an Age of Uncertainty* (Polity Press, Cambridge, 2006)
9. J.B. Bury, *The Idea of Progress. An Inquiry into Its Origin and Growth* (Macmillan, London, 1920)
10. R. Delort, F. Walter, *Histoire de l'environnement Européen, préface de Jacques Le Goff* (PUF, Paris, 2001)
11. M. Augè, *Futur* (Bollati Boringhieri, Torino, 2012)
12. E. Boncinelli, U. Galimberti, *E ora? La dimensione umana e le sfide della scienza, con Giovanni Maria Pace* (Einaudi, Torino, 2000)
13. L. Mumford, *Art and Technics* (Oxford University Press, Oxford, 1952)
14. Francis (2015). Encyclical Letter Laudato Si' of the Holy Father Francis on Care for Our Common Home. Vatican City: Vatican Press.
15. G. Ravasi, *Breve storia dell'anima* (Arnoldo Mondadori Oscar Saggi, Milano, 2009)
16. J. Le Goff, The configuration of European personality, in *History of Humanity – Vol. IV: From the Seventh to the Sixteenth Century*, ed. By M.A. Al-Bakhit, L. Bazin, S.M. Cissoko, A.A. Kamel (Routledge, Paris, 2000)
17. J. Le Goff, Prefazione, in *Histoire de l'environnement Européen, Préface de Jacques Le Goff*, ed. By R. Delort, F. Walter (PUF, Paris, 2002)
18. J. Le Goff, Costruzione e distruzione della città murata, in *La città e le mura*, ed. By C. De Seta, J. Le Goff (Editori Laterza, Roma-Bari, 1989)
19. J. Burckhardt, *Die Kultur der Renaissance in Italien* (Ein Versuch, Basel, 1860)
20. G. Simmel, The philosophy of landscape. Theory. Culture Soc. **24**(7–8), 20––29 (2007)
21. W. Dilthey, *L'analisi dell'uomo e l'intuizione della natura, dal rinascimento al secolo XVIII*, vol I (Nuova Italia, Sancasciano Val di Pesa, 1927), p. 15
22. Y.N. Harari, *Da animali a dei, breve storia dell'umanità* (Bompiani, Milano, 2014)
23. H. Jonas, *Dem bösen Ende näher, Gespräche über das Verhältnis des Menschen zur Natur* (Suhrkamp, Frankfurt, 1993)
24. L. Ramond, De la végétation sur les montagnes. *Annales du Muséum d'histoire naturelle*, anno IV. Quoted in Delort and Walter (2001), p. 404
25. J. Brunhes, *La Géographie Humaine: Essai de Classification Positive, Principes et Exemples* (Alcan, Paris, 1910)
26. Nobel Prize, The Nobel prize in chemistry 1995. 1995. https://www.nobelprize.org/prizes/chemistry/1995/press-release/. Accessed 22 May 2019
27. P.J. Crutzen, *Benvenuti nell'Antropocene! L'uomo ha cambiato il clima, la Terra entra in una nuova era* (Mondadori, Milano, 2005)
28. Earth Overshoot Day, Past earth overshoot days (2019). https://www.overshootday.org/newsroom/past-earth-overshoot-days/. Accessed 27 May 2019
29. WWF, *Living Planet Report - 2018: Aiming Higher* (WWF, Gland, 2018)
30. G. Bologna, Dall'economia della crescita all'economia della sostenibilità, in *Prosperità senza crescita. Economia per il pianeta reale*, ed. By T. Jackson (Edizioni Ambiente, Milano, 2011), pp. 17–46
31. H.E. Daly, *Introduction to T. Jackson. Prosperity Without Growth: Economics for a Finite Planet* (Earthscan/Routledge, London, 2009)
32. T. Jackson, *Prosperity Without Growth* (Earthscan, New York, 2009)
33. T. Piketty, E. Saez, *Income Inequality in the United States, 1913–2002* (University of California, Berkeley, 2004)
34. T. Piketty, *Le capital au XXI siècle* (Éditions du Seuil, Paris, 2013)
35. J. Perkins, *Confessions of an Economic Hit Man* (Berrett-Koehler Publishers, San Francisco, 2004)
36. R.G. Wilkinson, T. Pickett, *The Spirit Level. Why More Equal Societies Almost Always Do Better* (Allen Lane, London, 2009)
37. G. Anders, *The Outdatedness of Human Beings*, vol 2 (C.H. Beck Publisher, Munich, 1980)
38. N. Harari, *Sapiens. A Brief History of Humankind* (Harper, New York, 2015)

39. H. Jonas, *Philosophical Essays: From Ancient Creed to Technological Man* (Prentice Hall, Englewood Cliffs, 1974)
40. Á. Heller, *A Philosophy of Morals* (Basil Blackwell, Oxford/Boston, 1990)
41. W.F. Ruddiman, The early anthropogenic hypothesis: challenges and responses. Rev. Geophys. **45**(4), 37 (2007)
42. J.P. Crutzen, E.F. Stoermer, The "anthropocene". Global Change Newslett. **41**, 17 (2000)

Chapter 2
Sharing as Cultural Revolution

2.1 Communitarian Environment

The basic problem that we must escape, as men and as designers, is the risk of moving toward a reality that denies the beauty of humanity. The drying up of the perception of the environment's vital beauty is the first step in the domain of the techno-scientific apparatus, which led to environmental problems. We must stop this tendency, reestablishing a healthy relationship with the natural environment, and at the same time, we cannot allow the same depletion to occur in humans' spirit. This is the primary task we must face. The situation is well described by Federico Faggin,[1] one of the inventors of the "chip," who a few years ago, interviewed [1] in Vicenza about his challenge to build a computer that can learn on its own, criticized the contemporary scientific society as being guilty of spreading the idea that "everything is a machine.". According to Faggin, this idea led humankind to underestimate itself and to fall into a dangerous state in which machines can become our cages instead of our bridge to freedom. To quit this state, a post-digital neo-humanism is necessary. The example that Faggin used was in front of him: the Palladian Basilica, a World Heritage Site, designed by Andrea Palladio between the sixteenth and seventeenth centuries. A machine can photograph this wonderful work of art but cannot see it, and, even worse, cannot feel any emotions rising from its view. In a society governed by cold science, it is not just the inability to feel wonder that should frighten us, but also the inner emptiness that pervades modern individuals in this "liquid" society.[2]

---

[1] Born in 1941, after designing the first computers for Olivetti, "small" as a closet, after being one of the pioneers of Silicon Valley and having created the "silicon gate" that allowed him to build the first microchip, after founding Synaptics and having filed patents for the touchpad and touchscreen, the scientist from Vicenza has good reason to watch with amused detachment these digital philosophical "antics."

[2] In Bauman's formula for "liquidity," the reality of the postmodern is made up of inconsistency, contingency and mutability.

© Springer Nature Switzerland AG 2020
E. Giorgi, *The Co-Housing Phenomenon*, The Urban Book Series,
https://doi.org/10.1007/978-3-030-37097-8_2

The deafness of emotions, the unawareness of human actions, and the lack of authenticity are the bane of our society and our discipline's task will be to help wipe out this daily dullness and create the conditions for man to be able to return to shine and to do good.

2.1.1 Sharing

The beauty of humanity and environment is too often forgotten today. Economy and technology, which play a key role in defining the features of our times, find it difficult to speak about an intangible dimension that is too difficult to measure like beauty. The weight that the market and technology have placed on society has differed according to historical periods. So even individuals' needs changed along the historical changes. The physiological needs are basically similar in all societies, but the material requirements that refer to social and psychological skills vary significantly from society to society.[3]

One of the problems in our society is that the importance given to material commodities is excessive. In order to live in the society without being ashamed, we have to adjust to the level of material goods of others, continuing to push further up the demand for these types of goods. The big problem is that remaining in this paradigm does not allow us to find any way out: if being socially recognized and progress depend on material production, our hope of regaining an alliance with the environment will inevitably not be fulfilled. According to Tim Jackson,[4] this power of the social logic over the standard of individual life is closely linked to a steady weakening of the sense of the common initiative, which is also one of the victims of the *consumer society*.

This sense of individuality, linked with the struggle to have more and more goods for the for the sake of appearance in society, does nothing but worsen the problem, increasing competitiveness among individuals and creating anxiety and continuous expectations that reduce personal well-being. In other words, the meaning of life is close to becoming invisible. There is no doubt, for the good of society, the environment and ultimately for our own sake, this vicious circle of consumerism should be stopped with an efficient proposal, so that the structures of our society can be redefined in a new perspective of sustainable and lasting development. On the one hand, we should rethink our efforts in view of a public good in which we are called to participate; on the other hand, it is easier to change our own lifestyle if we are surrounded by a community that makes the same effort by acting in the same direction.

To break the vicious circle in which to be recognized in society we must necessarily consume, we should follow Gandhi's famous advice: "*live simply so that*

---

[3] Amartya Sen (b. 1933), Nobel Prize for economics in 1989.

[4] 4 Tim Jackson (b. 1957), Professor of Sustainable Development at the University of Surrey (England).

others may simply live," inviting people to change their lifestyle, approaching simplicity voluntarily. So, we can start to understand that the solution to the environmental crisis we are experiencing is in shared aims and in common goods.

We can see this on a small scale, for example, in some communities. In creating a restricted environment in which trying these new approaches has led to the creation of various typologies of "community," some of them have attitudes geared more to the protection of the environment, while other, more spiritual people want to recover a lost sense of spirituality.

These forms of community will be analyzed in detail later because of their importance as social "laboratories." These new fields, which can affect both the domestic environment and community, are valid opportunities to experience a new lifestyle that combines respect for the environment, welfare and new relational dimensions. This relationship between the environment, social relations, and individual well-being is the basis of all these ideas. In the last decade or so, we saw the creation of a global network coordinating the initiatives of these movements and communities, which see the solution to contemporary problems in the sharing of common goods.

However, sharing should not be understood just as a solution for environmental and social problems: we must also be aware that sharing, which involves social interaction, is a major human component. And in its being natural, it can reinvigorate our relationship with the environment that is now slowly fading.

Since ancient times the awareness of sociality, as an important component of human nature, has been present in philosophical and religious texts. Aristotle stated the importance of family bonds and social relations for individual well-being. Humankind is by nature drawn to sociality, so that our behaviors are closely linked to emotions caused by love, kindness, or gratitude (whether because of their presence or their absence) [2]. Moreover, there is plenty of scientific support for the empirical idea that relates social bonds to positive benefits, both psychological and physical.

As for involvement and volunteering in a community, Dolan, Peasgood and White [3] report that some studies report a relationship between SWB (subjective well-being) and membership in (non-church) organizations. Pichler's [4] analysis of the European Social Survey found that membership of more organizations increases life satisfaction. Helliwell [5] found that both individual involvement in non-church organizations and the national average membership of non-church organizations are significantly positively related to life satisfaction in his analysis of 49 countries from the WVS. However, from some studies, different results emerge, which indicate nonsignificant relations between civic participation and life satisfaction [6]. So we can observe a substantial correlation between happiness (SWB) and participation in communitarian life, even if these researches exhort always taking precautions, case by case. Reading these articles, an important consideration comes to mind: you can find the basic rules that allow high levels of perceived well-being (including the sense of community) to be achieved, but in many cases, they depend on the conditions of the context of belonging.

We saw two aspects of sharing: on the one hand, it is configured as a solution to regain a renewed relationship with the environment; on the other, the direct benefits gained from it seem to be clear. If, then, we only had the courage to change the paradigm that binds us to this type of society, this revolution would not only be productive, but also enjoyable. So far, we have seen the problematic aspects correlated with technological development, but the occasions on which these technologies allow us to create networking and sharing are becoming more frequent, making it possible to exchange and offer everything from services to goods.

The important fact that should be noted is that this tendency to share produces several benefits for the society: on the one hand, we can share what is not used or is little used; on the other, you can use a product or service without having to buy a new good. For Arun Sundararajan, professor at New York University, individual economy, occupation and environment will benefit because the investment opportunities are increasing for everyone, including for those with low starting capital, particularly the new technologies that enable manufacturing at low costs, and so anyone can become a venture capitalist.[5]

The fields in which these kinds of initiatives are developing vary greatly. Here we try to offer some examples of initiatives that are successful thanks to network sharing. We are dealing with a small, but important, paradigm shift: do not look anymore at "ownership," but at "use." So, we can save both money and environmental damages. It is possible to create credible alternatives to the current paradigm and be happy in a less materialistic way. Shared objectives must be the focus on which we concentrate; we must give new life to the concept of "public good."

2.1.2 The Common Good

By sharing we perform actions that become a common reality, a space, an activity or simply a good. Common goods can be divided into two distinct groups: those that refer to the goods of subsistence, necessary for life, and the global ones, such as climate, peace and all those goods coming from a collective creation. For both the categories, what is interesting to underline is the fact that the reference to the term "public good" not only denotes the good but also the ability of the society to carry it out and to enjoy the results. It is, therefore, also considered the relational dimension, the immaterial and affective world linked to our relationship with others. This double aspect, tangible and intangible, makes the argument closely linked with the architectural practice, so that it becomes the instrument that can realize the junction point between the reality of the places of the city and the immateriality of the bonds among the individuals who experience them, made up of stories, legends, culture, sensibility, and *genius loci*. Also, the American political economist Elinor Ostrom,

---

[5] Arun Sundararajan, professor at New York University in the "Science of Management," author of *The Sharing Economy: The End of Employment and the Rise of Crowd-Based Capitalism*, in an interview with Report, the Italian independent journalistic TV programme (RAI3), on May 24, 2015.

who dedicated her research to the value of sharing, confirms the role of architecture in creating community and allowing people to feel psychologically guaranteed and to be active in their daily life. This is demonstrable because, according to the American researcher, many communities are able to take care of common goods without government or private intervention, from the Swiss grazing pastures to some Japanese forests and the irrigation systems in Spain and the Philippines, explored in depth in the book *Governing the Commons.*[6] Some principles, such as the direct participation of citizens in the management of these collective goods and the absence of too fast technological and social changes, allow the public good to be ensured. Inghilleri, in analyzing the thoughts of Ostrom, explains the concept of commons as a form of social capital, which is based on true and honest relations, confidence but also shared values. It is this idea of social capital that merges together the members of a community and that allows them to act together toward a common direction [7]. "Social capital" is, then, the collective value of social networks and the inclination, which derives from them, to interact and to help each other. In the development of this capital it is clear that an important role is played by places in the city that allow them to meet, that revitalize the network of microeconomic activities and that compact the social environments of a neighborhood. Regardless of the public or private sectors, those elements that are able to create relational exchanges and meet the desires of the population can be defined as *common goods*, and it is here that architecture is called to allow people to feel psychologically guaranteed, allowing an identity and ensuring that we do not feel lost in society. They are, essentially, the membership of a family group, the sense of being part of a community and the possibility of self-realization. The architecture, together with the society, which must ensure accurate social securities, can and must contribute to promoting the establishment of these guarantees and this function can be performed in different forms, even in slums. Often overlooked and considered as problematic realities to be solved (later we will deal with the role of the *favelas* in creating community), slums can become examples to be taken as a reference in talking about relationships, social exchanges, and attachment to a place and the community. The difficult reality of the context requires counting on each other, so that the richness of the social capital is indisputable. Architecture must be at the forefront, then, paying attention to the needs of citizens and to ensure that people and places can fully express their potential and realize their wishes, in a shared way.

2.1.3 Community

Cavalli and Levi

"Community is a social group that generally lives in a definite territory and it is part of a society of which it incorporates many features but on a smaller scale, with a less

---

[6]Elinor Ostrom (1933–2012), American political scientist and winner of the Nobel Prize for economics in 2009.

complex and differentiated game of relationships and interests. Unlike the society in a community, there is greater emphasis on all-absorbing and sympathetic aspects of interpersonal relationships" [8]. In this definition accepted in contemporary sociological research, the Italian sociologist Alessandro Cavalli emphasizes the local character of the concept of community, in which there is a widespread feeling of common belonging and, sometimes, personalistic or solidary constraints, which are typical of ancient communities founded on agricultural production. The great changes in production processes, trade and transport have endangered these traditional forms of community. In fact, in this definition of community, Cavalli shows how several factors, including primarily the productive revolutions and globalization, contributed to breaking the previous community relations and to introducing individualistic and rational constraints in the modern society.

On the other hand, in giving the definition of "political community" Lucio Levi[7] underlines a relationship, in a quantitative dimension, between production methods and social relations—in particular, the evolution of the production methods used to deepen social relations, bringing the political community up to higher levels [9]. Unlike Alessandro Cavalli, Levi accepts, for community, higher levels, up to the continental one, but it should be noted that both of them connect the concept of community and human relationships with productive methods. In fact, for Levi, it is the social division of labor that institutes the way in which individuals develop their relations and the dimension of the groups.

Tonnies and Weber
In analyzing the meaning of community, it is even more interesting to deal with the problem of the definition through the concepts developed recently in sociology and philosophy, in particular by Tonnies and Weber. The concept of *community* incorporates the confrontation between society and community itself, as developed by the German sociologist Ferdinand Tonnies (in his book *Gemeinschaft und Gesellschaft*, edited in 1887). In this regard, Alessandro Cavalli specifies that Tonnies proposes a dichotomous conception of the term "social" between an intimate, instinctive, and trust-based world (*Gemeinschaft*, the community) and a public, conventional, and artificial world (*Gesellschaft*, society). For Tonnies, this community and society are two ways to develop relations that are diametrically opposed. In a society, each individual looks for their own individual goals, in a rational way and without considering the other actors' scopes. In contrast, in the community-based social relationship, the solidarity among the actors is the main characteristic. Moreover, Tonnies is convinced that in this last case, these relations are not voluntarily established, but come from spontaneous attitudes toward a "widespread nature" [8].

This dichotomous contraposition between the terms "Vergesellschaftung" (social condition) and "Vergemeinschaftung" (community condition) disappears with Max Weber, who, in his *Wirtschaft und Gesellschaft* (*Economy and Society*), uses them to indicate in a systemic approach two different processes and analytical

---

[7] Lucio Levi (b. 1938) is Professor of Political Science and Comparative Politics at the University of Torino, Italy.

2.1 Communitarian Environment

characteristics, with the addition of the concept of "Kampf" (conflict). So a shared system of values defines the community and differentiates it from the society, which, if the common approach to the same value system fails, generates conflict.

Victor Turner

The figure of a beneficial community, although not antithetical to society, but rather a "simultaneous presence" that accompanies individuals throughout their lives and defines their personality, strengthened in the 1960s with the Scottish anthropologist Victor Turner. In his thought, community (*communitas*) goes beyond conventional social ties and shows the characteristics of belonging and solidarity [10].

At this point, we can indicate a few firm points on our route: there is no pure community and no pure society and the need to make a distinction in this sense is not suggested by observation but by aspiration for an ideal. These two concepts were transformed to assume in contemporary sociology the distinction between "local" and "cosmopolitan" social relations. It is a purely descriptive distinction between behaviors related to a restricted community in which you live and behaviors oriented to a wider society [11]. Moreover, a social system like ours, based on mobility, is not compatible today with the old conception: the romantic *blood and soil* community is no longer feasible.

The sense of belonging to a group is, however, critical to the welfare of the individual. Sociality is part of our human being because it is only with this membership, more or less large and tangible, that we can recognize ourselves as individuals. It is not a coincidence that in Abraham Maslow's "classification of needs" two categories of needs out of five arise from social relations. In fact, after the basic needs that affect the physiological level and security, Maslow suggests the emotional needs (family ties, friendship, love) and estimation (realization and respect). For Harari, family and community impact more on the individual's happiness than money and health. Strong families and supportive communities contribute to the happiness [12]. We cannot accept losing the relational dimension and we know that the individual, in order to satisfy their needs and feel at ease, always looks for a community structure to belong to, which today is to a very large extent represented by virtual social networks. Some argue that these networks contributed massively to the dissolution of the community; for others, they occupied the space left by the sense of community, acting as a surrogate. Certainly, they have become an important tool for recreating virtual communities, in which the individual can satisfy, albeit in part, their relational needs. The purpose of architecture then becomes evident: to allow a revival of the sense of community beyond the new virtual forms. After all, architecture has always designed spaces that respond to the needs and characteristics of a society. For Edoardo Narne, Professor of Architectural and Urban Composition at the University of Padua, sharing and sociality are always the subject of architectural design research: all around the world, all the cultures of housing have searched for solidarity and sharing inside and outside the dwelling, looking for relations with the neighborhood and trying to avoid loneliness and isolation [13].

But can a regenerated architectural design create a new community for contemporary society? Can the creation of new spaces and the reformulation of a pact for cohabitation satisfy our new relational needs, without any imposition?

These are the key questions on the role that design must take in renewing the spirit of the community and allowing the individual to regain their social dignity.

2.1.4 Industrial Community by Adriano Olivetti

In recent decades, numerous reactions have developed to changes that have radically changed our idea of community, not only driven by political ideals or rebellions, although they certainly had a greater resonance because of these ideals. We can consider the season of protest that broke out in the USA with the "flower children," which then struck Europe in 1968 with protests and the birth of "municipalities," a phenomenon of response from the community that has persisted, despite losing strength and consistency until today. Perhaps because of the resonance these responses have made, even today the ideas of community and sharing are wrongly associated with ideological motivations or political affiliations. A change of attitude toward homologation and a disengagement of thought in recent years is evident. The deep crisis described at the beginning of this research has partly contributed to the collapse of certainties in our society, which is based on production and consumption. The environmental, social, and economic crisis has caused a rethinking process, so that in an individualistic society like the one we created, we now consider some principles of community sharing that in some cases have proved to be successful in our cities. Therefore, sharing can overcome the limits of family or work ties, creating new levels of community, sometimes never seen before.

In a world surely governed by forces of a global nature that affect everyday lives, we are seeing a slight change in course toward more human and communitarian dimensions of living. This small-scale field links with a second, of a much greater scale, which is precisely that of a unified world. Contemporary, or "globalized," man becomes aware of living in a unified world or in the "network society."[8] The coexistence of these two community realities recalls and testifies to the term "glocalization," which indicates precisely the relationship between local and global peculiar to the contemporary society, by combining the planetary connection, daughter of the information revolution and economic development, with the traditional and cultural dimensions of local contexts. This local dimension, along with his cultural background, must be rediscovered and valued. A slight parochialism is an enhancement of the human dimension of living and, therefore, it is positive.

Speaking about the role of smart cities in the modern world, Evgeny Morozov writes: "The 'smart village' should not even aspire to be the 'global village' of Marshall McLuhan: parochialism, in small doses, can be positive. A village should be proud of its local culture. A community feeling is important: it is not the

---

[8] Network society is a term coined by the Spanish sociologist Manuel Castells Oliván in the last decade of the twentieth century, referring to the changes caused mainly by information and communications technology (ICT).

2.1 Communitarian Environment

electronic house net imagined by Alvin Toffler in 1980 in his bestseller *The Third Wave*" [14].

In 1908, in Ivrea, Camillo Olivetti founded the first Italian typewriter factory, which in the 1930s became, thanks to Adriano Olivetti, one of the most important Italian industries and, after World War II, one of the most important in global information technology. Adriano Olivetti founded his industrial approach on the idea of using the industrial profit for investments in favor of the community. The fortune of the company, combined with Olivetti's ideas and his attention to culture, allowed the development of Ivrea, which is interesting from the standpoint of planning and community. One of the most important architectural works, the result of the approach of Olivetti and the design genius of the architects Gabetti and Isola, is the Olivetti Residential Centre in Ivrea: a residential amphitheater, where the residents can contemplate nature. Even outside the town of Ivrea, however, to strengthen the bonds between peasants and workers, the idea of community proposed by Olivetti was particularly successful. In 1945, Olivetti published *The Political Order of the Community*, which will constitute the theoretical basis on which the community project will be based, but which will also be taken up by Altiero Spinelli who, during his exile in Switzerland, met Olivetti. For Olivetti, a more conscious, civilized leadership, morally and materially healthy too, of local units would bring a renewed vigor to the entire state. Therefore, he proposed an experiment for a new way of doing politics, in which the forces of autonomy replace the central government to solve the main problems of the community. On the grounds of the 1945 manifesto, Olivetti founded 3 years later the *Community Movement*, which was a great success and led in 1958 to him being elected to the Italian Parliament. According to the idea of Olivetti, the term "*comunità*" reaffirms the idea of the movement to join, to cooperate, to teach and to build, creating a social environment capable, in the final instance, of "discovering vocations"[9] [15]. With the due considerations arising from timing, cultural and technological differences, we can say that the case of Olivetti is one of the most important references for this work. In his theoretical work and then in realization, Olivetti links the cultural aspect with the communitarian one in an alliance that allows man to develop, fleeing from the danger of an atomizing society and the loss of identity, which he defined as the greatest danger. The idea of Olivetti, as the main outlet, has a political dimension at the national level, while our analysis focuses on the role of the community regarding the individual and the environment at a local level. However, it is also true that we are interested in the importance that the communitarian dimension assumes to redeem man from a society that minimizes him, and this is the same objective as that of Olivetti. It is equally true that, beyond the politics, the ultimate goal is the same. In fact, Olivetti saw in these communities the basic unit of a great democratic project for the realization of a people's state, based on "small homelands," with at their center scientific, social, and esthetic

---

[9]An interesting excerpt from Olivetti's explanation of the term "communità" refers to Olivetti's idea of promoting a cohesive, free and tolerant community, following the motto that inspired the Church in the Council of Trent: "In the necessary things unity, in doubtful freedom, in all things tolerance."

values and where economics and technology contribute to the wellness of humankind. All this is coherent with the matter described above, which introduces our goal: a deep change that, starting from the culture, removes humankind from the yoke of enslavement to economic and technological logic. For Olivetti, the spiritual dimension is infinitely more important than any economic and political order, so it should become a central problem for contemporary democracies [15]. The entrepreneur of Ivrea points out the size a community must be that is neither too small to prevent the development of the individual and the community itself, nor too big as a contemporary metropolis that depersonalizes man and where a unity of interests and spirit is missing. Olivetti proposes a middle course, in small towns with an intense life, with harmony and peace, far away from the overcrowded metropolis, from isolation and loneliness [15].

Clearly, Olivetti, in formulating his idea of community, has in mind at least one case in which democracy, labor, and culture combine and interact to enrich the community, namely the Carl Zeiss Company in Jena (Germany), also quoted in *Il cammino della comunità*. Here, the city and the community represented the democratic order, with a unique control between university and factory. While the factory aided the university, the university helped the factory, in a circle that enriched the community of Jena [15]. In this context, highlights Olivetti, the labor of workers has a different value, because, if integrated in a system of fruitful development and mutual help, it serves a noble purpose and does not become a burden to the human spirit.

But Olivetti, deeply convinced of the revolutionary virtues of culture that give an individual their true power and true expression, reminds us that all the efforts promoted to change the society will be lost if education, human spirit, culture, and the light of intellect will not transmit to anyone [15].

2.2 Creating a Human Environment

In the twentieth century the architectural experimentation has undoubtedly seen a significant development, especially in the field of housing, from the district scale to buildings and apartments; with modernism, design was clearly inspired by the mechanical essence of machines, then, with postmodernism by the concept of the past and, finally, as noted by Nan Ellin,[10] by the networks, both as an ecological model, both as a computational model [16]. We are not going to compile here a compendium of projects, researches, and criticism that have occupied the architectural history of the last century (William Morris, Henry van de Velde, Adolf Loos, Peter Behrens, Walter Grophius, etc.), but we try to propose, through individual

---

[10] Nan Ellin (b. 1959) is an urban designer and Dean of the College of Architecture and Planning at the University of Colorado Denver.

2.2 Creating a Human Environment

experiences or episodes, what was developed by the discipline in this "Short Twentieth Century."[11]

In the following discussion, we intend the term *design*, both in an architectural and an urbanistic sense, where *planning* is about the design of the territory and *programming* relates specifically to territorial economy. This elaborate meaning arises from the strong tie between architectonic practice and urbanistic planning, due to the common operational method applying the same rationalizing structure to both scales. Moreover, there is also the usual procedure of many architects to support design activity with administrative activities, in order to get direct access to authorities that work out urbanistic plans and decide on rules and guidelines regarding building development. Here are some cases. Jacobus Oud was appointed municipal architect in Rotterdam from 1918 to 1933; Adolf Loos worked in 1920 for the housing office in Vienna; Bruno Taut was a housing councilman in Magdeburg (1921–1924). In 1925, Ernst May became town-planning consultant in Frankfurt, and from 1925, Martin Wagner was the first architect in Berlin.

2.2.1 Rationality

Using a famous quotation by Robin Evans,[12] the architecture of the last centuries has produced a "general lobotomy performed on the whole society" [17]. Having decided to meet the needs of the technological apparatus, architecture has lost its way for the environment. It reduced or even removed smells, noises, sounds, sights, differences, and confusions, reducing daily life to a sequence of private events. It was a reduction that led to the spread of architectures and spaces that were indifferent and insignificant without any link with the territory in which they arose. As Richard Sennett underlines, a city is no longer made up of flesh and stones, but of stones and denied bodies [18]. Certainly, the role played by technology was decisive, as it not only had an impact in terms of innovation in building methods, but also a responsibility to change the way of thinking about the project and beyond.

As the Italian art historian Eugenio Battisti underlines in the book *Architettura, ideologia e scienza*, technology affects architecture since it leads us to apply precepts established by the "economic-productive structure" and collected in rules and handbooks [19]. Battisti also highlights an element of contradiction that interests us, particularly in demonstrating how the revolution produced from applying strong rationalizing tools to architectural design has caused upheavals but not necessarily improvements in the lives of inhabitants. Rationalist practice has betrayed the

---

[11] "Short Twentieth Century" is a well-known expression used by the historian Eric Hobsbawm in the book *The Age of Extremes* to describe social changes that occurred in the period 1914–1991 and the failure of several ways of predict (and imaging) the future (fascism, communism, capitalism, liberalism, socialism, etc.).

[12] Robin Evans (1944–1993) was an architect and historian. He widely lectured at renewed universities in the UK and the USA.

original intent for which architects had introduced it into their design action and showed its true nature (dominated by technology) by drying up architectural activity. Le Corbusier himself, who with a rationalist approach created memorable works (as the convent of La Tourette), at a certain point wonders if the people who inhabit his works, thought as machines for living, will really find a human dimension. Also, during the construction of the Unité d'habitation in Marseille (1947–1952), Le Corbusier had doubts about the project. In fact, Andre Wogenscky, an architect who worked with him, reports his concerns: "Am I not wrong? Will the inhabitants of the Unité of Marseille be happy? Would you like to live in the Ville Radieuse?"[13] The architectural value of this work is, and will always remain, unquestionable for the housing ideas reflected here. However, in analyzing the results, we can find in Le Corbusier's doubts the awareness of some strong limitations of the proposed city model. The iterated system adopted for the construction of the city of the future does not convince. In fact, in the following decades, his followers, who were less sensitive, realized all around the world whole neighborhoods and superstructures of inhuman dimensions, based on the repetition of the same housing cells, uninhabitable places, obstacles to socialization where situations of maximum degradation quickly developed.

The human dimension of living, that of communities, villages and sharing, was sacrificed in favor of utopian visions. This subordination of architecture to the technology apparatus becomes evident with the story of Matera, when in the 1950s the old city was considered uninhabitable on the grounds of the researches of some American sociologists. Therefore, authorities built from scratch a new town for 20,000 inhabitants on a nearby plain that was anonymous and insignificant. A city built in the rock was abandoned, the result of layering and living experience of a secular stone civilization that had always coexisted with nature and in nature. A place where people lived the natural man/environment relationship was abandoned in favor of modern apartment blocks, which, even if ecologically sustainable, are certainly unable to reproduce the spatial relationships, the cultural and economic aspects of the ancient city. These modern attempts, which certainly arise from ambitions to reduce the environmental impact, have failed to integrate the new projects with existing systems, thereby removing choices, expressions and vitality for existing communities. In recent years the environmental sustainability of architectural interventions became an essential element in the evaluation of a project. Too often, in any case, design choices centered on the satisfaction of energy-saving requirements neglect other aspects of sustainability important to the inhabitants and environment well-being.

The apparatus of technology, as Paul Valery affirms in *Eupalinos or the architect*, caused the abandonment of "singing" buildings for new "dumb" ones. A similar fate befell the Riviera Ligure, which Calvino says was marred by "concrete fever," which raised thousands of identical tenements. Regardless of the place where they are situated, their history and green surroundings, these buildings rise white for six to eight floors with a lot of windows and balconies overlooking the sea, from Genoa to San Remo [21].

---

[13] The words of Le Corbusier [20].

2.2 Creating a Human Environment

A similar fate awaited cities where urban planning and zoning generated a fragmented city without continuity, "without soul or character." The result is cities, districts or parts of cities deprived of resilience, which in a city is located, by nature, in the possibility of interaction with the population. A mere technical design shortens the time dimension of a living place, on the one hand by speeding up its realization, and on the other hand, by shortening its usage time. Instead of arising and evolving over decades, if not centuries, and continuing to evolve under the pressures of the changes, new districts are calculated and created straight off, managed and abandoned or demolished when something external and unforeseen in the program happens. So even the city becomes a consumer good, with planned obsolescence, adapting well to the industrial world. Among the cold calculations of planning and programming, elaborated against a backdrop of globalization and standardization that is now ubiquitous, not only the city pace is forgotten, but also that of man and his spaces. In the development that Frampton calls "megalopolitan," isolated buildings, skyscrapers, and rows of houses, side by side but not integrated with the new infrastructures, create an urban landscape that is lonely and without identity.

In 1972, the authorities began the breaking up of Pruitt-Igoe, a 33-building complex—a redemption dream for the people in a city like St. Louis where the racial problem was very strong. Pruitt-Igoe was presented as a symbol of hope for a new life of the community. That hope soon turned into a nightmare, so the administration of St. Louis was obliged to demolish the whole area, which had been completed just 18 years before by the well-known architect Minoru Yamasaki, who designed the World Trade Center towers.

This mega-urban intervention, an icon of a ruinous design theory, was the subject of numerous films and documentaries, including *Koyaanisqatsi* (1982), a documentary by Godfrey Reggio on contemporary society, which through a four-minute sequence tells the story of Pruitt-Igoe, from abandonment to demolition; and *The Pruitt-Igoe Myth*, a 2011 documentary, which describes in detail the whole story, from hope to disappointment and destruction. With the demolition in 1972, for the first time in the entire history of architecture, an extensive construction of a prestigious and recognized designer was destroyed with dynamite, signifying with extreme violence that a detached and depersonalizing way of realizing houses is no longer viable and that human living must be reconsidered from other points of view. Modern architecture had built, on that occasion, an environment that was unsuitable and hostile to socialization. Since that moment, in the USA, the message has been clear: no more house-containers for a huge number of families, but rather smaller groups where social relations can be cultivated.

2.2.2 Architecture and Soul

What, then, should the new approach to architectural practice be? Should it still be based only on technological goals, as it has been mainly in recent decades? The proposal presented here should not be considered an attack against the technological aspects of design, which we know to be the basis of today's practice. Our thesis is that the technological features must be at the service of a nobler aim, the alliance between humanity and environment. The architecture must follow the path drawn poetically by Antoine de Saint-Exupéry: "I do not reject the stairs of the conquests that allow man to climb higher. But I have not confused the means with the purpose, the staircase with the temple. It is essential that the stairway allows access to the temple, otherwise it will remain deserted. But only the temple is important. It is urgent for man to find the means to grow around him, but it is only the stairway that leads to the man. The soul that I will build will be the cathedral, because this alone is important."[14]

Besides those episodes of denouncement against an architecture that has taken a direction opposed to man, we recall, here, some examples in which culture and nature can integrate. They are designs in which the attention to a technological system never triumphs over the search for the collective and human dimension of living. After all, Niemyer is right: "Architecture is just an excuse. Life is important, man is important, this strange animal who owns soul and emotions, and the desire for justice and beauty" [22]—a justice and beauty that become real in a feeling of harmony between people and nature. And, as Nan Ellin observes, anthropologists and cultural theorists are more and more willing to consider culture as a part of nature and not the reverse [16].

Professor Zheng Shiling explains the guidelines for the project of the Chinese Expo of 2010 as a search for an ideal city, reflected in the conception of a harmonious city—a harmony that, according to the Chinese traditional philosophy, is a balanced relation between peoples, nature, bodies, and minds [23], a harmony that has been present for millennia in the Chinese culture and today in the design approach too. Zheng Shiling emphasizes with the ancient words of Confucius that morality refers to the relations between individuals, society, and nature, revealing the importance of an ethical reference in the relation between man and environment (both social and physical). In other words, this ethical approach in urban and architectural design is a fundamental aspect of a good neighborhood and it also has a role in reasoning on urbanism. Nan Ellin, in his *Slash City*, evidences how modern architecture requires a constant control, completeness, and full planning. The result is a sequence of tall monofunctional buildings linked by highways, totally unrelated to other parts of the territory and with very high sustainability costs. Instead, a so-called "integral urbanism" looks for connections, communications, and sociality. This become real with the increasing attention given to poly-cultivation and mixed uses. The importance of the local level is increasing regarding the production and consumption processes and also the networks between the productive, financial and institutional actors.

Now we take into account the architectural philosophy and practice of two architects, who are not Western but are linked with the Western tradition. The first one fought in his country against the fashion of Western modern architecture and the second one considers his country architecture to be a victim of the same drying tendency as the Western one.

---

[14]Antoine de Saint-Exupéry (1900–1944) in his work *Citadelle*, quoted in [15].

2.2 Creating a Human Environment

Hassan Fathy

In the same period as when Le Corbusier was building his "Unité d'Habitation" in Marseille (inaugurated in 1952) and the architectural community was taking his "machine" work as reference for a modern architecture, in Egypt the Egyptian Department of Antiquities called for tenders in an urban project. The aim was to design some new villages that would replace, for touristic reasons, other old ones and rehouse the 7000 inhabitants of Gourna, near the Tomb of Nobles. The proposals were basically of two typologies. The main one was based on a "machine" approach, with repeated units contained in enormous concrete buildings capable of hosting all of the 7000 people moving from Gourna. Hassan Fathy's project introduced a construction methodology based on traditional materials and building techniques, incorporating the benefits of modern know-how with a vernacular style. Thanks to this philosophical approach coming from a humanistic connection between people and their own environment, Fathy created ecologically and economically sustainable buildings, totally integrated with the village community. The importance of the project was not just the result of using traditional knowledge and materials to create the built environment, but was also due to the spatial urban design, which was capable of enhancing neighborhood relations thanks to the sharing of open spaces. In this project it is possible to perceive how a rediscovered tradition and historical references can help us think critically of the benefits of technological and economic development.

Charles Correa

The same attitude toward the human scale and the attention to collectivity are also present in Charles Correa's work, in which we can see again the attention to the psychological dimension and the traditional approach. In fact, Correa denounces the situation into which architectural design has fallen, based on economy of scales and technological development. It is his contemporary generation of architects that lost the basic principles of an architecture founded on the ingenious and innovative people's culture. The strengthening of the human relations and solutions that would really be culturally sustainable are the basic principles of his architectural design. Without escaping from the present and looking for an ideal golden past, we should be able to recognize the contemporary challenges and formulate proposals for a better future. It is not a matter of neglecting or using technologies, but rather of using them in the best way. We must find a way to combine technology with a new appreciation of spirituality, simplicity, and slowness, without forgetting the always present metaphysical peculiarities of places.

2.2.3 Inhabitants Live with Soul

We understood that if people who live in spaces see their traditions respected, their sense of territorial belonging, their history, in a word their culture, then the "architecture for users" will cease and people will return to being "inhabitants" again, with all the dignity that this word brings. Once we have defined, as we have just seen with the examples of Fathy and Correa, the cultural approach that the project must have for the architecture of the soul, we can propose a reflection on the use that the inhabitants must be able to make of these spaces: people must be free to improve their spaces. In this perspective, architects and designers maintain their specific function. They must be able to design in harmony with the place and with people by understanding their needs, leaving the project free to fit with the environment and man in the best way. The changing circumstances, discussions, beauty, and movement should not be dismissed but rather they are the elements from which the project must start, they are the essence of the environment in which man lives. We are architects and we must remember that beauty, friendship, and dignity are the ingredients that make man live.

Therefore, we must design spaces that can recover and support the innovative and living forces of popular culture and tradition. Our duty is to design flexible spaces that can adapt to the will of the inhabitants—living spaces that will not be completed by the architect's design, but by the disposition of the people in a dynamic process. It would be impossible to design spaces neglecting the values and dreams people have.

The Italian architect Boeri added as a warning that we will have to avoid the danger of any ghettoization, designing to allow people to legitimize their independence and free them from alienation and anxiety, and today this is only possible by creating networks and interchanges [24].

There are several examples of situations in which the inhabitants of a place have the freedom to partly modify their environment to create realities in which they can feel more comfortable. Let us think, for example, of how, on Tenerife island, the canals called "*barrancos*" are used. These big canals are the results of several lava flows and became the natural drainage system on the island. When there are big storms, they become real filled creeks that collect all the rainwater from the northern side of the Teide Vulcan (the tallest mountain in Spain) and bring it to the seaside. During the dry period, on the other hand, these canals, which are in front of houses, are occupied by the inhabitants who organize them as meeting places, with semi-public spaces, children's playgrounds and temporary furniture. This soft urbanization of the canals is carried out by the inhabitants themselves, using the stones brought by the floods and homemade material with which they create elements and spaces for sociality, from simple seats to playgrounds, even though they know that these structures will often be destroyed. This can be considered a suitable case of free spatial interpretation by the inhabitants.

2.2.4 The "Principle of Responsibility" in Architecture

Sometimes, when the rules imposed by architecture are too strong, people that still have a living culture find their right to act as inhabitants and citizens by breaking the rules and organizing their own living space. Just as the vitality, dynamism, and

2.2 Creating a Human Environment

beauty of childhood can disappear under the oppression of adult duties, in the same way the features of a city can disappear if oppressed by overly strong planning.

In all these situations where the constraints imposed by technological design are too stringent, human living is manifested through the ability to occupy and organize places for life. There are thousands of examples of this phenomenon of spontaneous interpretation whereby dumbly designed places are made to sing by people (no longer users). Just think of Mediterranean streets, of the gardens and squares of the South American *favelas*, of the occupied buildings in the cities of South-East Asia. All these examples show how the inhabitants, if they remain residents and nonusers, can create much better places than those designed for them but with other aims.

We really believe in the words of Prince Myshkin in *The Idiot*, "Beauty will save the world," which, in that context, have a very mystical meaning. According to Salvatore Settis, we cannot use those words like a mantra because "beauty won't save us if we can't save beauty" [25]. Following the viewpoint of Settis, we can add that our society must conform its fundamental requests to a value system based on the protection of nature, of human health and of public ethical and individual morality. This principle of public interest recalls old themes such as the *"commune bonum"* or the *"utilitas publica,"* which were concepts of "public interest" present in the charters and in the statutes of Italian *comuni* (municipalities) and pre-unitary states. Among their objectives there was also the protection of the rights of future generations on the cities and the landscape, now threatened.

These principles were accepted, for example, in the Constitution of Bolivia, promulgated in February 2009, which recognizes the right of every citizen to "a healthy, protected and balanced environment." Moreover, "[t]he exercise of this right must be granted to individuals and collectives of present and future generations" (Article 33). Salvatore Settis recalls here the already quoted "responsibility principle" (1979) of Hans Jonas, who underlined how the common destinies of man and nature, now in danger, force us to rediscover the dignity of nature and to preserve its integrity [25].

Civil Commitment Becomes a Duty

Urban constraints and market laws are destroying those faculties that have enabled man to create spaces in which he lives. This capacity, however, has not yet been completely extinguished; in situations where the restrictions imposed by planning are too strong, or totally lacking, it manifests itself through the ability to occupy and organize space (with results often more on a human scale compared to those designed spaces).

There are a lot of examples of this phenomenon of spontaneous interpretation, by which silent places are made expressive by people. We can take as an example the David Tower in Caracas, a winner, at the Biennale of Venice 2012, as a model of "common space," an example of "collective and informal living." It was developed by the urban think tank (Alfredo Brillenbourg, Hubert Klumpner) and Justin McGuirk, and in particular by the inhabitants of Caracas and their families who created a new community and a new house starting from an abandoned and unfinished building. This experience can be considered as a model of squatting: a model of "common ground." The architects were able to recognize the potentialities of this project of transformation: a spontaneous community that created a new identity occupying, with ability, the David Tower. This example can be considered an inspiring model for informal associations.

Often if we consider a "planned" city, we will have a perception of rigidity, monotony, difficulties in orientation and a lack of public squares and open spaces. On the other hand, if we refer to a nonformal urban space, a space not designed with rules or following regulations, they will appear complex, with open organic and functional spaces. In this kind of urban environment, we can say that the networks become a reality.

It was in accordance with these ideas that in 2009 the group of Christian Kerez accepted the invitation of the Secretaria Municipal de Habitacao (SEHAB) of São Paulo (Brazil) to present some proposals for the improvement of the living conditions in these kinds of nonformal urban spaces. They started by comparing two basic and opposed strategies: the modern German quarter of "Gropiusstadt" accomplished in 1960 by Gropius and Paraisopolis, the second largest *favela* in São Paolo, located in the southern part of the Brazilian metropolis. The comparison conducted by the architects and proposed in *Abitare* [26] points out how the economic activities, the flexibility, and the capacity for development of an urban model like the *favela* can be more relevant than the modern model. *Favelas* have their specific urban qualities, made up of differentiated and narrow urban spaces, where people meet and children play. In this way, people who have to move from difficult situations can find in this kind of residential system better conditions of ventilation, light, and safety, while maintaining their link with the vernacular architecture and vitality of the *favelas*. "This project pays respect to the inhabitants of the *favelas* who are mostly proud to live in their community, as most people are proud of their own roots to which they can relate" [27].

2.3 Five Architectural Ways to Create a Community

The followers of the *Neues Bauen* conceived individual buildings as representatives of a social collectivity in relation to technology and the tools of production. This would be, then, the way to understand the Bauhaus. In 1922, the critic Adolf Behne wrote "a responsible technology driven by the awareness and realized through a collectively interconnected work, leading especially to the realization of a deep and mutual dependence and to a 'conditionality of the relativity' … organizes the mass. And from this mass the community crystallizes" [28].

Architecture, however, not only expresses the social community but also has the power to form social relationships. Purpose and usefulness are real esthetic properties of architecture: the architect, who manages to achieve efficiency and utility, becomes a creator of ethical and social character. In this case, people who use the building will be inspired, through the structure of the house, to behave better in their mutual relations. Architecture thus becomes a creator of new social rituals, but, of course, architecture is not the only force that enables humankind to fulfil itself in its social and relational dimension. In fact, as Christopher Alexander underlines, "cities and buildings can become alive only if they are created by a community of people who share a common language model."[15] Alexander comes, then, to distinguish between "dead" and "alive" architecture, one in which no relationships are created and another where meetings occur and where social relationships are born. However, a consideration remains, which summarizes the thoughts expressed by Hertzberger [28]: as architects, we can create spaces that have the potential for both individual and social use, but we do not have the power to determine the outcome. In other words, we can create the conditions in which the relationships occur, but we cannot monitor their success. There are, at least, five ways in which the organization of space can contribute to creating communities. The following considerations aim to introduce these five ways, expressing a general thought focused on the aspect of design and intervention. Of course, each of these topics deserves an explanation that would require several books and here is just a quick introduction, mainly for those who are less familiar with the topics.

2.3.1 Commons

There is a pluralistic interpretation of this term, which can identify existing, or potential, elements on the different scales of design. They can be common resources (natural and urban), common spaces[16] or interventions focused on the enhancement of communities. In this last case, the one we are more interested in, commons are all the interventions that emphasize one of the fundamental roles of design (perhaps considered marginal in recent decades), that is, the improvement of the social and economic conditions of a local reality, reached through the involvement of people during the design process and through the building of a community awareness and a cultural process of emancipation [31]. Solutions designed to solve problems, often very tangible, in difficult conditions and using local techniques and traditional materials can look like a re-proposing of the concept of "architecture without architects," but this is the real architecture.

If we think differently, it is just because the architecture of the last decades left a wrong picture of our mission. We do not look just for the shape or for formal solutions, but we try to solve the real problems between people and the environment in which they live. This is certainly not a contrast to the modernist architecture, as is a traditionalist architecture, but rather it is an *ethnographic postmodernity* that gives us a less pessimistic view of the collective future, constituted by increasingly intense symbolic exchanges and by contacts between different identities. In the *commons*,

---

[15] The words of Christopher Alexander quoted in [28].

[16] The wide framework "City as a Commons" aims to read the urban settlements as a network of common goods and shared resources, capable of revitalizing social fabrics and increasing the sense of belonging [29, 30].

we do not look for the contemplation of the form, of the skin or of the construction suggestions, but rather we look for the beauty in reference to the places and to the people who live architectures, in alliance with their environment.

Usually, this kind of intervention has an aim that mainly focuses on disadvantaged areas (often rural on peri-urban areas), where social problems are more relevant. In particular, attention is addressed to the *favelas* that are now seen not just as places where architectural intervention is needed to resolve technical or social problems, but also as places where architecture can take references for theoretical ideas and design, in terms of esthetic and, above all, of social and community life.

This kind of approach cannot be based only on the evaluation of a society and a culture performed by the architect. In other words, it cannot be an imposed architecture but, rather, the result of a continuous negotiation between the meanings proposed by the designers and those by the people, future inhabitants of the designed spaces. Dialog and discussion should not be perceived as a complication of the design process, but as a necessary encounter that makes architectural work not only a sculptural form, but an ethical work.

Since this is a relationship between problematic realities and designers, who were often trained in the schools of Western architecture, a comparison with colonialism may be legitimate. While with the colonialist criteria, established design criteria were transported to the new reality, in the architecture of the "commons," architects carry with them an important technological and cultural know-how that is applied to the context and to the single situations in which they act. It is also important, however, out of respect for the seriousness of our work and of the situations in which we act, to emphasize that commons cannot be considered just from a moralistic or performative point of view. It is a mistake to think that ethically based architecture cannot have the same rights to beauty as other kinds of architecture, or that the social focus of ethically based architecture could distract from the research of esthetical purposes. The esthetic/spatial value, even if with limited resources, can and must be achieved, not least because beauty is a key component of human well-being. Combining two design features becomes important: on the one hand, aspects of inviolability that give protection, and on the other, aspects of flexibility that can guarantee the ability to adapt to changes and to developments.

2.3.2 *Public Spaces*

Public space is the element that, more than any other, characterizes a city. Here people can meet as citizens, with their rituals and their customs, and here, according to features that change from city to city, the public space expresses the way in which people live and perceive the city. A famous quote by Joan Clos, executive director of UN-Habitat, affirms that it is its public spaces and not its private spaces that contribute to defining the character of a city. The public space, moreover, always keeps a function of debate that can hardly be recreated in other realities, physical or not: it is the nature of public space within contemporary societies, according to Salvatore

2.3 Five Architectural Ways to Create a Community

Veca, that influences the "*utopie al plurale*" (utopias in the plural), the possible worlds—the public space as a place where ideas can still be expressed and debated, where you can look for consensus and adhesion. Veca underlines the role that public space has in allowing the society to express its freedom, because it is the laboratory of diversity, of experiments, of alternatives, "of the non-compliance to the rules and of the variety of social identities" [32]. In contemporary cities, public space becomes the place where a variety of "possible worlds" takes shape. The public space is, therefore, the social, not the institutional, place to express pluralism.

However, the public physical space is undergoing major changes and is losing, in part, its monopoly of place for debate and organization of social actions. In fact, as noted by John Parkinson in *Democracy and Public Spaces*, the phenomenon of a public life that moves away from the reality of public spaces is not something recent [33]. For some decades, at least, the public sphere of political debate and social initiatives has moved away from the built spaces of squares and markets, moving first to the newspapers and to the broadcast media. Now, even these tools are overtaken by a revolution that made the public sphere digital and accessible through the desktop, far from any physical reality. From manifestations and protests coordinated via Facebook and Twitter, or from the collections of signatures that take place on web platforms like "Change.org," to sensitize public opinion and leaders, we can say that activism and its communication pass, in some way, through the network.

It is a fact that, through smartphones and tablets, we can access information and exchange opinions, without any distinction regarding the part of the world and the context or the physical space where we are, which can lead to thinking that it would be useless to reflect on designing public spaces. However, it is because the function of the public space changes that the way we design it must also change. Thinking that we can put a stop to the use of modern communication tools would be wrong, perhaps even harmful. We must, however, be able to combine these instruments with a renewed model of public space that offers to man a functional potential, such as to make it once again a place full of life and character for the city. The rekindled interest in the public space cannot be achieved just through spectacular, visually striking interventions that act on open spaces. [34].

We believe that the public space can still create community and be a focal point within the city. The project "Project for Public Spaces" (PPS) [35] shares our vision on the problems and on the need to rethink the design of contemporary space, and proposes strategies through which the public space of cities can be revitalized, bringing new life and a new human dimension to the citizens and to the city. It means working with the existing resources, creating a system with the buildings, a more varied functional program, linked with the proposals of the municipality and having the courage to experiment, with an approach that can be considered "lighter, quicker and cheaper."[17] As the PPS project shows, rethinking the infrastructure network can also give new meaning to urban spaces. In fact, we should not underestimate

---

[17] Nonprofit planning, design and educational organization dedicated to helping people create and sustain public spaces that build stronger communities. The PPS was founded in 1975 to expand on the work of William (Holly) Whyte, author of *The Social Life of Small Urban Spaces*.

the fact that speed counteracts the living in and belonging to a space. The speed and the almost unlimited presence of cars in the streets curb the use of spaces like roads (squares have become parking lots), which, given slow mobility, could function as neighborhood "public space," for district meetings or for occasional chats between neighbors. Rethinking the infrastructure system means working on different elements of the design (attractions, destinations, identities, borders, etc.) to return neighbors, waterfronts, and parks to people. These new open spaces can bring new lifestyles: eco-friendly mobility, a sense of identity and, above all, new relationships between people and their environment. Another aspect that is often highlighted is the relationship between public space and human activities. The value of a dynamic place is greater than the sum of its parts. In the dynamism and vitality[18] of the spaces is hidden their true value, human but also economic. The value of a place cannot be understood just in the context of investments and profits, but also in the aspect of opportunities for citizens—opportunities like public spaces where human interactions, networks, and connections are formed; like cities in which the interests and the abilities of people meet the economies and the opportunities to create business from the bottom. Citizens become the entities that give a shape to the economy, bringing it back to a human level, linked to the territory and local production.

Thus, rethinking the public space is not limited just to formalist talks about space, but goes further by bringing to light the link between public space, public good, and human environment.

2.3.3 Icons

Engraved in humankind's imagination or schematized in a logo, the buildings of official and institutional power and the monuments of a heroic past represent the most famous icons in which communities can recognize themselves. Through the transmission of the cultural heritage and the creation of those spaces that make a frame for our daily rituals, architecture becomes a key player in the creation of imagined communities.

So strong is the ability of architecture to create a sense of belonging that it is used not only in its physical dimension but also through its representation. Therefore, the European Union has printed banknotes representing architectures of the European cultural background and which are well fixed in the mind of every European. These structures, which do not really exist in any part of the continent, are abstractions of buildings that strongly recall clear images of the architectural environment that Europeans have in mind.

Moreover, architecture can also become an iconic mirror of its contemporary society: Bauman [36, 37] shows how the approaches to the design of residences in a society governed by the fear of others and by skepticism toward being different is

---

[18] What Alexander called "architettura viva" [28].

configured through non-flamboyance or intimidation, showing the features of a fortification.

Additionally, architecture also has the distinctive feature of extending the collective memory of persons or events even beyond the reminiscence of those who directly knew them. Again, as Forty observes, the construction of commemorative buildings is one of the oldest architectural functions [28].

Furthermore, architecture has the incredible power to evoke, with its symbols, messages that are deeply studied in cases of landmark projects and, often, the meanings raised by the icon-architectures are subject of long considerations and strong negotiations. These considerations can easily go beyond the issues related to buildings to get talking about art and monuments. We can think of the heated discussions to develop the Vietnam Veterans Memorial in Washington, a project developed in the complicated context of disputes about the Asian war.

According to Carmen Popescu, professor at the Sorbonne in Paris, identity and identification are caused mainly by time and space, which are nothing more than fundamental coordinates of architecture [38]. The temporal dimension links with the size of tradition and history, while the spatial dimension refers to territory, city, and environment. Therefore, architecture assumes a key role in the processes of collective identification (local or national) and becomes an even more important element if we consider the recognition of belonging to certain ethnic communities. So it is possible to consider architecture as a tool to create a sense of belonging, but also a sense of modernity, as in the Brazilian case with the union between "traditional" and a "sense of modernity," through the architecture of Lucio Costa, Lina Bo Bardi, and Oscar Niemeyer.

The images and meanings that architecture gives us are, indeed, carried by some mental relationships that we have and that are derived from memory. The relationship between memory and architecture is an issue that is very important and difficult to treat. Ruskin spoke of architecture as a tool in the hands of man that allows him to conquer his own forgetfulness more easily than poetry. With architecture, in effect, not only is the thinking transmitted but also the physical essence of what the ancestors have touched, seen and shaped. For Ruskin, in fact, architecture is the human art that, more than any other, allows the formation of a sense of belonging to a community: through collective and shared memories, architecture creates identity [39].[19] With modernist architecture, however, this attitude changes. Preceded by Nietzsche's essay "On the Advantage and Disadvantage of History for Life," the modernist behavior toward memory is outlined in his total removal from the architectural debate. The rationalist principle conformed men to standard users, whose emotions and feelings had to be removed, in the same way as the concept of history, memory, and identity had to be ignored.

It is precisely the absence of these themes in the modernist thinking that provides further evidence of how the evocative power of architecture and its ability to create

---

[19] The Poetry of Architecture is a collection of articles published in London's architectural magazine under the pen name of Kata Physin ("According to Nature") in the biennium 1837–1838.

identity are key aspects in bringing man, community and feelings to the center of design. Memory is not only important because it allows us to create images of the community and to identify with them, but also because, as Forty reminds us quoting Walter Benjamin, it allows humankind to resist the supremacy of history.[20]

Reintroduced in the architectural debate, the memory becomes the "conscience of the city": it is the reading in the present that we give of the past and that allows an understanding of the city. With its architectures and permanencies (and with their mental images that we create), the city enables us to define a collective memory, allowing us to keep together the community itself. It is when architecture is not able to give voice to the memory and to the permanencies that this awareness fails, as is happening in contemporary societies, where alienation and loneliness are taking over.

Finally, there is a last point of view to consider, which refers to the authors Paul Connerton[21] and Dolores Hayden,[22] who see in the physical architecture not the creator of icons and collective memory, but only the theater of common rituals and ceremonies that are the final originators of shared identity [40, 41].

2.3.4 Co-working

The revolution in communication tools and the transition from print to digital media, which supports a connectivity that is becoming faster and faster, more powerful, portable, and affordable for everyone, have imposed an important change on the working systems. Today, and even more in the future, physical and digital spaces will contend the world of work, workplaces will be more and more independent of time and geographic location, and jobs based on knowledge will lead the market [42].

Additionally, if we think of the crisis affecting the Western economy for more than a decade, we understand that a large part of the business world had to evolve (or should evolve as soon as possible) in trying to survive. The limits of industries, market, and business are going to dissolve, and open systems and networks will characterize the next organization of work. It is enough to look at today's debate to understand that the concepts of community, sharing and networking have become the keywords of the working world. Hi-tech innovation will certainly bring advantages, especially for the capital and the elites, but it will create inequalities even deeper than the current ones and unthinkable uncertainties in the jobs field. The rationalization of production processes in the Industrial Revolution led to the

---

[20] According to Benjamin, history is a science of the nineteenth century that creates an event's narration controlled by predominant powers [28].

[21] Paul Connerton (1940–2019) was an anthropologist and a reference scholar in his discipline for studies on the concept of "memory."

[22] Dolores Hayden (b. 1945) is a professor emerita of architecture and urbanism at Yale University and founder in Los Angeles of the group "The Power of Place."

2.3 Five Architectural Ways to Create a Community

disappearance of the concept of *homo faber*, so this new hi-tech production method will lead to the definitive assertion of automation at the expense of traditional occupation [43]; but work will still be central in the new economy: it does not contribute just to the sustenance of individuals, but gives meaning to their participation in the life of society [44]. Frank Duffy, founder of DEGW, the international architectural and design practice best known for office design and workplace strategy, underlines how the new working relations impose quitting the "box" structure and entering a system of interaction, talking and working collaboratively [42]. So the narrow bi-univocal connection that is established between work and the social dimension of individuals is understandable: work to create the social dimension and sociability to allow the work to develop and enhance, creating networks of relationships.

The envisioning of the workplace for the coming years cannot overlook the combination of people, technology and spaces [45]. So, in changing the working methods, the spaces housing the new activities must clearly also change. Our task, therefore, is to understand how to manage these changes. In the last century, a debate calling for rethinking the places and the ways of working, to preserve a human dimension, in a reality that, by contrast, tended to crush the individual, was always present.[23] The workplaces are becoming more and more undefined: they are often a combination of physical spaces and virtual platforms; people can be working in the headquarters or in their own homes; wireless technology allows that open spaces, without fixed positions, become the contemporary offices. The main companies came to incentivize external options based on their employees' schedules (with the aim of reducing the number of workplaces and consequently the space needed for the company's business).

The relation with the urban context will also change: the local community will be involved in the company's activities, demolishing the walls of the old workplaces and opening the social and economic flows of opportunities present in the context and in the society. Co-working is a typical example of this opening and one of the most interesting entrepreneurial phenomena, albeit often undervalued, of the last two decades.

Co-working is often debated with an ideological approach and with wrong interpretation models. So, we intend to present co-working as a form of collaboration and sharing, featuring many facets and engaging multiple disciplines. Basically, we can thin down co-working as a space where several individuals work, alone or in a group, on their own project, sharing the resources and the cost of a traditional office,

---

[23] In the last 1920s and 1930s, the sociologist Maurice Halbwachs (1877–1945) argued that changing job would result in a dangerous loss of the sense of belonging and that therefore it was necessary to rethink work and post-work environments to strengthen the social ties [28]. In the 1980s, Henry Mintzberg, an academic in management science, wrote *Structures in Fives: Designing Effective Organizations*, in which he not only proposes forgetting the Taylor schemes linked to the organization, but also considering the organization of work as the key component of socialization. This means placing, in its antithesis with the model of Taylor, equal trust in the abilities and in the contribution that each one can provide.

but generating a vibrant environment that promotes personal relationships, business opportunities, and the sharing of skills and experiences.

In fact, despite the common misunderstanding, co-working is focused mainly around the aim of creating a network. So, design is called to create adequate spaces to provide the best possible conditions for each co-worker to make their relations flourish. So, it is not only a matter of desks, but also of shared spaces for relational activities; for this reason, they are usually spaces that allow noncasual encounters, collaboration, and discussion. Therefore, if, on the one hand, co-working creates opportunities for collaborations and new business, on the other hand, it also fulfils the human need for sociality and social recognition (a problem that should not be underestimated in a context in which new technologies have pushed workers and employees to work alone from home).

If we were to indicate a date for the birth of contemporary co-working, this was in 2005 with the opening in California of the "Hat Factory" by the computer programmer Bred Neuberg. This was the first workspace open to other professionals who shared a passion for their job and who were interested in combining it with a form of innovative experience. Since 2005, the phenomenon of co-working has continued and has experienced exponential growth. More and more young people (the average age of 70% of co-workers is between 20 and 40) are looking for these innovative spaces and they often become the direct managers of them [46]. In 2006, the "number of co-working spaces worldwide" was 60, becoming growing to 600 in 2010, 5780 in 2014 and 37,000 in 2018. The same exponential growth was seen in the "number of people working in co-working spaces worldwide," which rose from 21,000 units in 2010 to 510,000 units in 2015 [45].

Another research, promoted by Coworking Resources [47], shows quite different numbers (related to the fact that the two researches consider different definitions of co-working) but similar trends. In particular, this last research proposes a forecast for 2019, which will be the first year in which the number of openings of new co-working spaces will be lower than the number for the previous year.[24] Anyway, across the world, the demand for co-working spaces continues to be high in the less populated states and countries. Moreover, this research underlines an aspect that is particularly relevant for the co-working market: while some years ago it was thought that major co-working brands could monopolize the market, the trends shown in the research display an upward trend in individually owned and independent spaces.

Obviously, each co-working has its own peculiarities in terms of space and offered services. What binds them is always, and in all cases, the idea that from sharing may arise experiences of collaboration and growth.

---

[24] The estimated number of openings in 2018 was 2188 worldwide and 991 in the USA; in 2019, the estimated number of openings was 1688 worldwide and 696 in the USA.

2.3.5 Contractual Communities

Mankind has always felt the need to gather together in communities and in common settlements. The ways in which these associations have taken shape and have been managed can vary considerably from case to case, depending on the historical moment, the main scope of the aggregation, the presence of more or less strong legal constraints and the wishes of the inhabitants.

Without paying too much attention to the research on the characteristics and differences of these types of community forms, we merely present and analyze below two classification proposals by Italian authors on this issue, and finally, we will give some consideration to the significant reality of a local contractual community. Having an overall view of contractual communities is important because within them we can find co-housing, which we will examine in detail in the later stages of the research. We must be aware that the following classifications are reductive since the organization of these ways of sharing life strictly depends on the history, the context, the internal organization, and the purposes of the communities [27]. Before starting this overview, we must bear in mind that the typologies of co-living promoted by functional needs (such as subleases or apartments rented to students) must not be considered because they lack in the social-ethical aim.

Classification by Brunetta and Moroni

Two Italian urban planners, Grazia Brunetta and Stefano Moroni, propose a classification of the contractual communities according to the form of ownership and some other characteristics[25] [48]. It is interesting to underline that often these communitarian realities are defined as "institutional experiments." In fact, in having to organize a social asset and to define some relational structures, these communities become real laboratories, where the institutional forms and relations are studied and elaborated in their more essential conditions. The main distinction is based on the two characteristics related to the form of ownership:

- *Contractual communities of owners*, such as the Home house associations, the Co-housing communities and the Contractual communities of tenants;
- *Contractual communities of co-owners (common lands)*, such as the Collective territorial propriety and the Residential cooperatives.

Brunetta and Moroni also describe some common features of this type of contractual community that allow them to be defined as such.

- *Private property*: All cases of contractual communities base their value on the idea of private property. By several political and social movements of the twentieth century, we are often led to consider all the community experiences as an overcoming (and denial) of the idea of private property. The consequence is that

---

[25] Grazia Brunetta is Professor in Urban and Regional Planning at the Polytechnic of Turin. Stefano Moroni is Professor in Urban Planning at the Polytechnic of Milan.

this misunderstanding remains alive and strongly affects a real-estate market in which the demand for housing is focused on private property.

- *Territoriality*: A community is clearly referable to a precise land or area, with well-defined territory and boundaries.
- *Willingness to participate*: The decision to participate in these forms of co-habitation must be the result of a free choice and not imposed by external forces, as in the case of government social housing projects, for example.
- *Social contract*: Nelson [49] explains how the social contract that occurs in the contractual communities is neither a hypothetical one, as John Rawls expresses in 1971 in his essay "*A theory of Justice*" [50], nor an implicit one as theorized by John Locke in 1690 [51]. For Nelson, the social contract that occurs in the contractual communities is a real social contract.
- *Reciprocity*: Within the community there is a relationship between the inhabitants that is based on mutual help and on reciprocity, as a general method of action. This does not mean, however, not acting with the same principle toward people outside the community. In fact, often the communities try to propose their lifestyle for the social realities in which they appear.
- *Rights of private citizens*: Internal communitarian rules are developed inside the community itself respecting private and public law.
- *Internal control and sanctions*: Within the communities defined rules must exist and help in managing the relations among the members and sanctions can even be planned.
- *Self-financing*: The project is realized thanks to the financial efforts of the community inhabitants who do not receive any kind of public funding. There are several cases, however, in which public intervention may be present initially with a public concession of land or a first project funding.
- *Contributions in terms of benefits*: Living within these communities implies the compromise of contributing to the life of the community and brings benefits in terms of well-being and support.
- *Services offered by the community*: The community usually offers several services, within the community and externally, alternatives to the existing ones, often reaching higher quality levels than those provided by public administrations.

The differences among contractual communitarian forms are those features that we have already seen before: they are basically related to the typology of owning (owners, renters, co-owners, etc.) and to the decision-making system (boards, consensus, etc.). In this differentiation the motivational and ideological components have low value.

Classification by Sapio

Antonella Sapio, a child neuropsychiatrist who devoted herself in particular to psychosociological studies, deals with the issue of co-housing as part of an analysis on the evolution and crisis of the nuclear family model, explaining some reasons why people choose forms of shared life. This evolution has led to the formulation of new housing typologies through a different redistribution of private and common spaces.

2.3 Five Architectural Ways to Create a Community

According to Sapio, these spaces, with a particular reference to co-housing, allowed new forms of sociality and new forms of living in solidarity [52]. According to the author, there are three main forms that we can classify communities as: (1) co-residences; (2) co-habitations; and (3) nonresidential forms [52].

- *Co-residences*: These are based on cooperation among families living in private dwellings, sharing, however, spaces, moments and activities in order to create environments with high social and relational standards and to humanize the relations between neighbors. The qualitative and quantitative level of sharing can split this category into two subcategories:
  - *Solidarity condominiums*: with shared space and common assets managed by the community that are relatively limited, and the appearance of the buildings is closely linked to that of traditional buildings;
  - *Co-housing*: next to the private apartments, there are shared spaces for different activities, which can vary widely according to the single case needs. At the architectural level, the presence of common spaces is often recognizable due to their dimension.

- *Co-habitations* or *eco-villages*: Wide complex housing facilities often placed outside the city and with a notable eco-naturalist footprint. This form, due to the choice of an alternative lifestyle and the willingness to propose a new idea of settlement, reflects more than any other the experience of the communes of the 1960s.
- *Nonresidential forms* or *territorial communities*: These represent inter-family/interpersonal networks, in which social relations develop between individuals who live in the same geographical area but not in the same residential complex. A lack of sharing space does not necessarily imply a low level of sharing, which can even stretch to having mutual funds. Local communities become important, both because of the increasing levels of interpersonal relationships and because they often offer to territories different opportunities in terms of services or activities.

A Territorial Community: The Federazione dei Sette Comuni
A very remarkable example of this kind of territorial community can be found in the Asiago plateau (Italy), the *Federazione dei Setti Comuni*, which can be translated into English as the *Federation of the Seven Municipalities*. The ancient meaning of this federation and its historical alliance with the environment and the landscape has been brushed up on in the last few decades by the literary works of Mario Rigoni Stern.[26]

The origins of this federation date back to the thirteenth century and it always maintained its independence with a federal government, the *Reggenza*, up to the

---

[26] Rigoni Stern (Asiago, 1921–2008) was an Italian writer, considered by Italo Calvino one of the best Italian writers. He wrote mainly about his experience during WWII and about the intimate relation between the individual and the environment.

beginning of the nineteenth century. People had a central role[27] and the land was shared among all. This is the feature that interests us: still today the territory presents a particular community organization with a wide management autonomy. The land is not the property of the state or of private individuals but that of the community, and so the actions on the territory, such as the cutting of trees, are carried out only in the common interest. A very famous quote by Mario Rigoni Stern states: "In the territory of the Seven Municipalities there are no castles of nobles, no villas of lords, no cathedrals of bishops, due to the simple fact that the land is of the people and its fruits belong to everybody, as is the ancient custom." Life in the *Federazione dei Sette Comuni* has always been based on an intimate relationship with the environment[28] and on the honest values of sharing, which seem to be missing in the contemporary society far removed from this possibility of life.

References

1. G.A. Stella, Il miraggio dell'anima in bit, la sfida sbagliata del digitale. Corriere della Sera (2014), p. 39
2. R.F. Baumeister, M.R. Leary, The need to belong: desire for interpersonal attachments as a fundamental human motivation. Psychol. Bull. **117**(3), 497–529 (1995)
3. P. Dolan, T. Peasgood, T. White, Do we really know what makes us happy? A review of the economic literature on the factors associated with subjective well-being. J. Econ. Psychol. **29**, 94–122 (2008)
4. F. Pichler, Subjective quality of life of young Europeans. Feeling happy but who knows why? Soc. Indic. Res. **75**(3), 419–444 (2006)
5. J.F. Helliwell, Well-being, social capital and public policy: what's new? Econ. Model. **20**(2), 331–360 (2003)
6. Y. Li, A. Pickles, M. Savage, Social capital and social trust in Britain. Eur. Soc. Rev. **212**, 109–123 (2005)
7. P. Inghilleri, Towards an architecture of the commons and of identity. Lotus Int. **153**, 48 (2014)
8. A. Cavalli, Comunità, in *Grande Dizionario Enciclopedico*, ed. By V. Aa, vol. V (Utet, Torino, 1986), pp. 490–491
9. L. Levi, Comunità politica, in *Dizionario di Politica*, ed. By N. Bobbio, N. Matteucci (Utet, Torino, 1976), pp. 181–183
10. C. Bordoni, La fine della comunità. La Lettura – Corriere della Sera (2014), p. 2
11. N. Abbagnano, Comunità, in *Dizionario di filosofia*, ed. By N. Abbagnano (UTET, Torino, 1993), p. 940
12. N. Harari, *Sapiens. A Brief History of Humankind* (Harper, New York, 2015)
13. E. Narne, Il Novecento: Un secolo di sperimentazioni sulla residenza tra visioni ideologiche, disastri annunciati e innovazioni tipologiche, in *L'abitare Condiviso: Le Residenze Collettive Dalle Origini al Cohousing*, ed. By E. Narne, S. Sfriso (Marsilio, Venezia, 2013), pp. 13–37

---

[27] The maxim of the Reggenza was: "The good of the people is the good of the Regency and the good of the Regency is the good of the people." In Italian: "Il bene del popolo è il bene della Reggenza e il bene della Reggenza è il bene del popolo."

[28] "Wood … must be cared for more than a vineyard. The wood needs to be cultivated, but the cultivation of the wood doesn't give a seasonal harvest, nor an annual one, it gives a secular harvest" [53].

References 57

14. E. Morozov, La città intelligente è un po' scema. La Lettura – Corriere della Sera (2014), p. 5
15. A. Olivetti, *Il Cammino Della Comunità (1959)* (Comunità Editrice, Ivrea, 2013)
16. N. Ellin, Slash city. Lotus Int. **110**, 58–72 (2001)
17. R. Evans, Figures, doors and passages. Archit. Des. **48**(4), 267–278 (1978)
18. R. Sennett, *Flesh and Stone: The Body and the City in Western Civilization* (Norton, New York, 1994)
19. E. Battisti, *Architettura, Ideologia e Scienza. Teoria e pratica nelle discipline di progetto* (Feltrinelli, Milano, 1975)
20. E. Narne, S. Sfriso, *L'abitare Condiviso: Le Residenze Collettive Dalle origini al Cohousing* (Marsilio, Venezia, 2013)
21. F. Cevasco, La profezia amara di Calvino. La Lettura – Corriere della Sera (2013). http://lettura.corriere.it/la-profezia-amara-di-calvino/. Accessed 14 March 2019
22. O. Niemeyer, *Il Mondo è Ingiusto* (Mondadori, Milano, 2012)
23. Z. Shiling, *Changing Shanghai, from Expo's After Use to New Green Towns* (Officina Edizioni, Roma, 2011)
24. C. Boeri, *Le dimensioni umane dell'abitazione. Appunti per una progettazione attenta alle esigenze fisiche e psichiche dell'uomo* (Franco Angeli Editore, Milano, 1981)
25. S. Settis, La bellezza ci salverà. La santa alleanza di ambiente, paesaggio e cultura. la Repubblica (2014), p. 49
26. C. Kerez, Nuovo quartiere di abitazioni. Porto Seguro Sao Paulo. Abitare **524**, 32–39 (2012)
27. A. Mariotto, Il cohousing come pratica di cittadinanza organizzata, in *Cohousing e Comunità Solidali*, ed. By Studio Tamassociati, Vivere Insieme (Altreconomia, Milano, 2012), pp. 17–32
28. A. Forty, *Parole e edifici. Un vocabolario per l'architettura moderna* (Pendragon, Bologna, 2004)
29. F. Sheila, I. Christian, The city as a commons. Yale Law Policy Rev. **34**, 281–349 (2016)
30. S. Stavrides, *Common Space: The City as Commons* (Zed Books, London, 2016)
31. P. Nicolin, Architecture meets people. Lotus Int. **145**, 12–15 (2011)
32. S. Veca, *La Gran Città del Genere Umano. Dieci Conversazioni Filosofiche* (Ugo Mursia Editore, Milano, 2014)
33. J.R. Parkinson, *Democracy and Public Space. The Physical Sites of Democratic Performance* (Oxford University Press, Oxford, 2012)
34. N. Bertolino, Spazio pubblico 2.0. Il progetto del paesaggio urbano dopo la rivoluzione digitale, in *Esercizi di Architettura*, ed. By A. Bugatti (Maggioli, Segrate, 2013)
35. Project for Public Spaces, Projects & Programs (2019). Retrieved from https://www.pps.org/projects. Accessed 5 March 2019
36. Z. Bauman, *Trust and Fear in the Cities. Proceedings of the Conference Trust and Fear in the Cities* (Bruno Mondadori, Milano, 2004)
37. Z. Bauman, *Liquid Times: Living in an Age of Uncertainty* (Polity Press, Cambridge, 2007)
38. C. Popescu, Space, time: identity. Nat. Identities **8**(3), 189–206 (2006)
39. J. Ruskin, *The Poetry of Architecture* (John Wiley & Son, New York, 1873), p. 71
40. P. Connerton, *How Societies Remember* (Cambridge University Press, Cambridge, 1989)
41. D. Hayde, *The Power of Place. Urban Landscape as Public History* (The MIT Press, Cambridge, 1995)
42. ARUP, *Living Workplace* (Arup Foresight, London, 2011)
43. A. Günther, *The Outdatedness of Human Beings*, vol 2 (C.H. Beck Publisher, Munich, 1980)
44. T. Jackson, *Prosperity Without Growth: Economics for a Finite Planet* (Earthscan/Routledge, London, 2009)
45. ARUP, *Living Workplace* (Arup Foresight, London, 2017)
46. R. Valentino, *Coworking Progress. Il Futuro è Arrivato* (Nomos Edizioni, Busto Arsizio, 2013)
47. K. Hobson, Global Coworking Growth Study 2019. Coworking Resources (2019). Retrieved from https://www.coworkingresources. Accessed 25 August 2019
48. G. Brunetta, S. Moroni, *La città intraprendente, comunità contrattuali e sussidiarietà orizzontale* (Carocci Editore, Roma, 2011)

49. R.H. Nelson, *Private Neighbourhoods and the Transformation of Local Government* (Urban Institute Press, Washington, 2005)
50. J. Rawls, *A Theory of Justice* (Harvard University Press, Cambridge, 1971)
51. J. Locke, *An Essay Concerning Human Understanding* (Clarendon Press, Oxford, 1975)
52. A. Sapio, *Famiglie, Reti Familiari e Cohousing* (Franco Angeli Editore, Milano, 2010)
53. M. Rigoni Stern, Non è la borsa che deve governare. Tuttolibri – La Stampa (2008), pp. i, iii

Chapter 3
Sharing as Cultural Preexistence

3.1 Learning from History

The action of a community developer, which was described at the end of the previous chapter, is a propriety intrinsic in architectural practice, known and used since the beginning of civilization, to protect a community and to create a sense of belonging. The history of architecture is full of a great number of projects whose purpose was to create a common good, to share spaces, to increase the safety of inhabitants, or to create identity. Reading a good book on the history of architecture, one of the clearer aspects that emerges is the recurring involvement of the population in activities that contribute to defining the built environment of societies: creating squares, markets, churches, palaces or urban walls, defining the whole images of the cities and even of the territories. By the word "involvement" is intended both (1) an active participation in projects, first of all with the practices of self-construction, but also participating in the "common" ventures of a city, such as the construction of urban walls; and (2) a passive participation, which refers to the capacity of the population to read the architecture, understanding its meanings and being able to take advantage of the spaces that architecture has created. While the first type is in some way coming back to be a current topic, related to the practice of participatory design, which can give new life to the architectural discipline, the second type seems lost. The human faculty of perceiving architecture and reading its meaning seems to have dried up.

One of the first Italian architects who pushed to restore the practice of people participation in design was Giancarlo De Carlo, one of the greatest exponents of the architectural discipline in the Italian panorama of the twentieth century and a member of Team X. In one of his essays, "About Participatory Design" [1], he presented the idea that organizing and giving shape to a space was a common heritage, since someone who built their own house knew very well what their needs were and had precise ideas on how the space had to be organized to respond to its practical needs,

© Springer Nature Switzerland AG 2020
E. Giorgi, *The Co-Housing Phenomenon*, The Urban Book Series,
https://doi.org/10.1007/978-3-030-37097-8_3

60   3 Sharing as Cultural Preexistence

and how it had to be configured to become its own depiction, so many people participated in a widespread culture of living.

From the Renaissance to the Enlightenment to the Industrial Revolution, this knowledge started to disappear, becoming the exclusive domain of the designer, who applied it according to the socioeconomic influences of the periods. Without any doubt, architecture and design must regain the vital energy to engage people again in the processes of transforming the environment, becoming deeply open to participation and making people involved again in their social and natural environment. This is not achieved only by asking or talking to people, but also by reading what the "daily life" and time have transcribed in the physical space of the city and of the territory.

Architecture has always been able to organize the spaces of a community and to clearly represent its identity—an identity defined through cultural representations, in which people can recognize themselves. For this reason we can say that architecture is one of the clearest cultural expressions of humanity[1] and that it represents, therefore, one of the most important tools for reproducing cultural representations and identity, both individually—as the development of one's own personality—and in terms of community and a sense of belonging.[2]

For all these reasons, this chapter presents some cases of historical communitarian experiences from four different continents, to show how this tendency to live together was present in societies all around the world. The proposed examples are residential cases that represent an infinite percentage of all international experiences, and the reader, certainly, will be able to make several comparisons with other cases that they already know. These selected projects have been chosen because they are considered, by the author, to be relevant architectural practices that create very clear and well-defined social environments, where inhabitants could develop a sense of identity and attachment to the community. Moreover, these cases show how this communitarian environment helps in enjoying the welfare of belonging and social recognition. These case studies have some features in common, for example: (1) they are expressions of a sense of community typical of their culture; (2) productive activities coexist with the residential function; (3) they are characterized by a free will regarding membership; and (4) the sharing of spaces and activities is achieved as a recognized need of the community. Before looking at them in detail, it should be pointed out that at the beginning some other examples of shared life were also considered, including monasteries, dormitories for students, and social housing (in particular with reference to the renowned case of Karl Marx-Hof of the Red Vienna). In the end, these projects have not been considered: although the

---

[1] In 1919, Walter Grophius defined architecture as the clear expression of the noblest thoughts of men, of their fervor, of their humanity, of their faith, and of their religion.

[2] In the case of the construction of the Tiburtino area, a neighborhood in Rome, through a project by Quaroni and Ridolfi, architecture also becomes a representation of society. The residential districts built by INA-Casa over the years after World War II are perhaps the only design outcome of the so-called "realistic architecture": informal and composed of a typological mix, more suitable for the rural villages that the people who would have been given new homes came from.

levels of sharing, referring to the spaces or to the activities of these groups, represent an important aspect of these realities, several of their features are not in line with the concept of "community" as we are working on in this book. In particular, for the monasteries and convents, they would certainly take on a significant importance if the analysis of the communitarian aspect was focused on the typology or the urban design. Moreover, they are not representations of a conventional way of living, as the relationship between the communitarian and private dimension is totally reversed compared to ordinary life. As regards the reasons for this phenomenon, Trabattoni [2] observes that since all the activities were held in common, the private space of the individual was relegated to the dormitories, generally organized in cells.

For the cases of the dormitories and social housing, their absence in the book is related to the fact that the social experience of living in environments with shared facilities is due to the condition of the residents, who often cannot afford a different kind of housing solution. Moreover, in the case of dormitories, the experience of sharing spaces and time is limited to a fixed period of time, imposed by the membership on the individuals at a college.

3.2 Europe

3.2.1 Polis

To explain the way in which Greeks experienced the relation with the idea of community, it is important to introduce some aspects regarding their way of understanding the world. In fact, their understanding of nature and of the relationships among its parts has significantly influenced the Greek society, even in its organizational and institutional structures. The way the Greek society is structured is very dependent on these aspects, in particular the idea that the ancient Greeks had of "community" and the physical output that this concept acquired in the form of cities or urban settlements. In the first chapter, we said that the Greek reality was very fragmented, so every small society (even a few kilometers distant from each other) draws up its own peculiarities. However, one aspect that unifies and governs all Greek *poleis* was the isonomy. This means that all the people considered citizens[3] of a city had to follow the law deduced from the conception of a universal order: that natural order that ruled the world, from nature to the cities. In 1949, Maurice Le Bel wrote about the *law* that governed the Greek cities, which can be considered as the "binding force of the city", which was born with the city itself and that every citizen had to follow, since it was a universal law [3]. Now, being a proven harmony between the forces of nature that are stable and remain constantly in balance and harmony in a circular

---

[3] We must hold tight to the Greek concept of citizenship. In particular, we refer to the social thought of Aristotle that, as Maurice Le Bel highlights, excluded slaves (actually the greatest number of people living in the cities) [3].

return of events, the same harmony would necessarily rule within the *polis*. Every citizen, therefore, felt part of the community in which they lived. It is interesting to note how every citizen found their own realization in their participation in the communitarian life and in the construction of the common good. In this regard, Reale, in "A history of Ancient Philosophy," notes the importance of *polis* in the moral value that covered the life of the ancient Greeks, who considered the city state to be the measure of moral life and who could entirely identify the individual with the citizenship [4].

However, it remains to underline how, for the development of the Greek *polis*, the appearance of the community and universal harmony was at the base of the essence of these unique realities that have allowed Western civilization to make such important progress. In this statement of "*La città antica: istituzioni, società e forme urbane*," Emanuele Greco recalls briefly the structural aspects of the Greek *polis*, highlighting this important point: the role of the communitarian ideology. A set of ideas, centered on the notion of "koinon-koinonia" ("common thing" or "community"), set out to the physical concept of "meson" (a shared space and common ground for discussion and debates), crystalizing inspirational principles of behavior into the system of the *polis*.

It is to these basic concepts that most of the ideas "intrinsic to *polis* formation" appear closely connected: that the territory and the population living there are a *common thing*, not a "private domain"; that at least part of the population should be included as co-participating in this *common thing*; that executive power must be exercised for defined periods and in rotation and that its exercise has to comply with the established rules and be subject to the *law*, to the *nomos*, a term that in its etymological roots refers to the notion of *sharing* [5]. Considering the social role of meals, which is one of the daily moments that better describe the culture of every society, we make another consideration.[4] In many Greek cities, including the famous Crete and Sparta, sharing meals became a key moment in the formation of the sense of community. Consequentially, the structure of the city and of its facilities also adjusted to this social phenomenon. Collective buildings, such as the *andreion*, were designed precisely for this: hosting communal meals called "*syssitia*." Emanuele Greco wrote that these buildings where the common meals, *syssitia*, were consumed were difficult to distinguish from the *andreion* of a private home. It should be underlined that so far it has not been possible to clarify the question of whether in every city there was only one *andreion* or if, at some time, in the cases of towns consisting of several residential areas, every neighborhood could have its own common building where meals could be shared. Another element coming from the social organization of the Greek society remains in doubt: whether the common meals of slaves and subordinates were held in the same building, or whether—which seems more reasonable—separate locales were provided.

---

[4]We will see that the sharing of meals is an extremely important element in forming a sense of community, including in contemporary co-housing.

3.2 Europe

Moreover, some more considerations on the results in terms of spatial and physical concepts are relevant. This concept of universal harmony, at the base of the society of the *polis*, was particularly supported by Heraclitus and by the Milesians. For this reason, a good reference to study could be the city of Miletus (Fig. 3.1), from which the majority of the representatives of this philosophical concept come. The plant at Miletus, in fact, has become extremely popular, especially because its

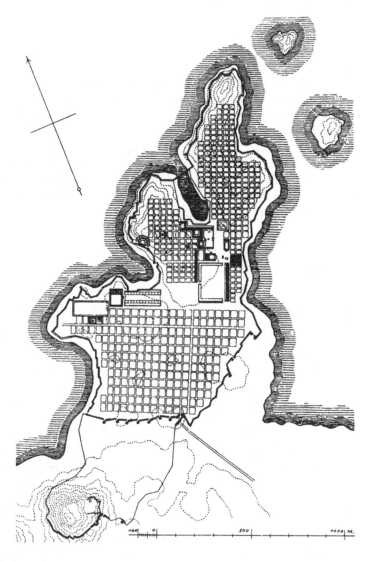

Fig. 3.1 Plan of Mileto, fifth century BC, here the role of public spaces in the definition of the urban morphology is very clear (https://it.wikipedia.org/wiki/File:Miletos_stadsplan_400.jpg)

design is attributed to Ippolito, who is taken as a reference by Greek urban planners.[5] That design we all know, however, is a hypothetical reconstruction proposed by archeologists in 1924 to underline the rationality of the Greek thinking [6]. Because of this forcing in the way of reading the urban design of this city, we analyze another coeval city: Pyrene, founded in 350 BC near Miletus. This city was founded on a strictly geometric design that leads the entire city to a chessboard of square blocks (35 × 47 m) delimited by streets of 3 and 4 m. The agora—an architecturally accomplished artifact—is inserted in this wire fence by removing a couple of blocks. From the *bouleuterion* to the *prytaneum*, from the archive to the court, from the *Stoa* to the *temenos* of Zeus, all these community buildings seem to be inscribed in the mesh of the houses, creating a sequence that spans the entire city. Rising up the hill, it started from the plan, from the peristyle of the big gymnasium, next to the stadium, to finish at the top with the theater and the minor gymnasium. So it seems strange that this sequence is not at all emphasized by major roads; instead it is arranged transversely. Even Pyrene is so structured: rather than around the street grid, around a system of common areas that unites the whole city. The fact remains that in Ephesus, birthplace of Heraclitus, the public spaces also took on such social importance as to be the spaces around which the design of the city developed. They connected with each other in a process that involved the whole city. Just as the essence of harmony that ruled the universe fell into crisis with the sophistry,[6] this conception of society and this strong spirit of citizenship also fell into crisis, and consequently this idea of urban design too.

3.2.2 Cascina

Since the eighth century, with the development of forms of territorial control such as vassalage, the European territory has begun to be dotted with production centers lost in the territory. This need to conquer the territory was dictated primarily by the quest to conquer new lands for agricultural production and exposed those who ventured into these areas to the dangers of an environment that was anything but hospitable. For Europeans, the European environment today seems far from being inhospitable: services, communication possibilities, and human presence have almost monopolized the European landscape. However, until a few centuries ago, this was not the case, and for those who lived in these places, the need for defense— against animals and humanity—was a priority. In particular, in northern Italy, a region where the typology of farms is typical, for centuries the raids of armies or bandits were not rare (Fig. 3.2). Thus, rural settlements had to play a protective role,

---

[5] Considered the main planner of the fifth century, the plants at Miletus, Thurii, Rhodes and Piraeus are attributed to him. He is considered famous not just for the orthogonal systems of his projects (which were already present in the Greek urbanism) but for a new reduced dimension of the blocks and streets, all with the same size.

[6] Protagoras (490–c. 420 BC) is one of the best-known sophists.

3.2 Europe   65

Fig. 3.2 Picture of the *cascina* Muggiano in Gattico (Italy) (author Alessandro Vecchi, licensed under the Creative Commons; https://it.wikipedia.org/wiki/File:Gattico_Muggiano.jpg)

often assuming the structure of small fortifications (and some of these settlements were later converted into real castles). In this context, the settlement of the farmhouse allowed a community of farmers to cultivate an area of the surrounding countryside, relying on shelter for work tools, animals and people. Often distant from other farms or villages, and even from the cities, these small settlements assumed the value of independent settlements. In this small reality of civilization in the midst of an adverse environment, a community was created based on the sharing of resources, spaces, activities, and common sense.

It should be noted that the phenomenon of farms covered a vast geographical area: technically the term *cascina* refers only to the building typology typical of northern Italy, but similar solutions can be found in almost all the flat areas of the Italian peninsula, and of many other European countries. Moreover, the phenomenon has evolved for many centuries, reaching maximum expansion between the seventeenth and eighteenth centuries. These two aspects meant that the phenomenon took very different forms in terms of spatial and social organization. We can, however, summarize some of the most relevant aspects for the discussion we need to do.

In fact, the basic typology of a farmhouse is a central courtyard defined by a system of buildings with different functions. In most cases, this rectangular courtyard was completely closed on all four sides to allow greater defense against external attacks. The buildings that defined it were substantial—those related to the residence (of the workers, the contractors and, sometimes, the owner), to the agricultural production with sheltering for the agricultural tools, to the stable and to the common activities. Depending on the size of the farms, these could contain buildings for public activities such as the school, a tavern, or more commonly a chapel or church. The size of the communities, in fact, could vary considerably between 4 and 20 families, all involved in productive activities [7]. Furthermore, the central space was usually left totally free from the presence of buildings or fixed elements, to allow complete freedom for productive activities that could require large open spaces for agricultural work (such as drying and beating of wheat) or for the celebration of events and parties. Basically, the idea of the *Cascina*, therefore, was a community created for defensive and productive reasons, where the people could enjoy a shared life.

3.2.3 Phalanstery

The Phalanstery, the most famous project of the architect Fourier,[7] did not find a direct realization, although it can be considered one of the utopian visions that, in some way, have seen at least an indirect and not pure realization. With Fourier being a visionary designer, even if the Phalanstery remains confined in the utopian sphere, it is a very interesting project since it represents well the design summary of a very articulate and idealistic thought. Firmly opposed to his contemporary society, which he considered immoral and based just on personal interests, he grouped his thoughts in Théorie des quatre mouvements, which was published anonymously in 1808 (Fig. 3.3). Among the bestiary of fantastic animals that he represents, the Phalanstery is the vision for the structure of a society of the future, which he describes in great detail. The exposed theory proposes seven stages that will enable the society to achieve harmony. After the current period, which is right between the fourth (barbarism) and the fifth period (civilization), a sixth del garantismo[8] and a seventh dell'armonia[9] will arrive. For both the societies, Fourier describes the ideal cities. The city of the sixth period is concentric and Fourier describes it in minute detail, from indicating the size of the roads and the urban indices, etc., that seem to anticipate the nineteenth-century building codes. The society of the seventh period, on the other hand, is described in the words of Servier as a society that will be changed by the passion carefully used by the social structures, where the workers will be grouped according to their dominant passion or inclination; but, at the same time, these groups will be in opposition to each other in order to create a healthy competition and competitive spirit [8]. In this society, everyone has the right to a minimum income to survive, regardless of the work, and the wage system is abolished, replaced by a sharing of profits, so that the workers could become the manager of

---

[7] François-Marie-Charles Fourier (1772–1837).

[8] In English, "defense of civil rights."

[9] In English, "harmony."

3.2 Europe

Fig. 3.3 Original drawing of the Phalanstère by Charles Fourier (1772–1837) (Biblioteca Nacional de España; this file is licensed under the Creative Commons Attribution-Share Alike 4.0 International license; https://commons.wikimedia.org/wiki/File:Fourier_o_sea_explanación _del_sistema_social_1870_d1.jpg)

their own work.[10] Fourier, in his *La théorie de l'unité universelle* of 1841, describes his idealistic project of the Falansterio as a building designed for an association of 1500–1600 people. Fourier also imagines his idea of the territory where the Falansterio could be set: a square alloy land will need a beautiful river, crossed by hills and suitable for varied crop fields and not far from a big city, but also not so near, to avoid intruders [9]. Thus the Phalanstery is a settlement created from scratch, avoiding any relationships with the actual context.[11] The Phalanstery is then organized by phalanges, smaller groups, which, as well as subdividing the society, subdivided the Phalanstery into areas reserved for different groups. The description continues, then, with the details of the project, accompanied also by a financial plan

---

[10] This is a very strong social vision that will be picked up by some communities in the Israeli kibbutz, or in some cases by intentional communities of co-housing and eco-villages.

[11] It is interesting to note that many of the utopian visions, from the *Sforzinda* by Francesco di Giorgio to the *City of the Sun* by Tommaso Campanella, prefer to confront "*tabula rasa*" contexts, like islands in the sea or places immersed in more or less vague plains—utopian visions that, to better express their concept of society, flee the existing reality, thereby evading that fundamental step that would indicate the processes of transformation of the society.

that imagines an initial investment divided by dividends between the residents of the community. These are perfectly free to join or not the community, without imposition or obligation of any kind.

Architecturally, the project includes a central building intended for public functions and public spaces called Seristeri (like the Tour d'Ordre with a clock, telegraph and carrier pigeons), while the two wings are laboratories and areas linked to production. The community building is organized through collective public spaces. The wings are served, for all their development, by a gallery that works as the main distribution (walkway) on the first floor. It allows there to be a covered link that would unite the whole complex. The gallery, three stories high, allows entry to common areas for recreation, and onto it face the rooms of the building, while the ground floor is left for warehouses, workshops, and driveways. As in most utopias, agriculture plays the character of the primary activity for man and the industry takes on, at least, a secondary role. On the other hand, almost 20% of the jobs are reserved for scientists and intellectuals. These wise and experienced persons, who represent the Academy, assisted by a group of elected bodies, were in charge of the major decisions.

Although the vision of Fourier was clearly an idea, he always tried to look for investors who were willing to fund his project, but without success. In fact, the Phalanstery never saw the light of day, even though it had many disciples around the world. In the USA there are many examples of communities that followed the principles of the Phalanstery, with the most famous, perhaps, being the experience of Phalanx, founded in 1843 in New Jersey, with 125 members, and that of West Roxbury, Massachusetts, in 1841. The most significant achievement, however, remains the French one by Jean-Baptiste Godin (1817–1889), a pupil of Fourier and enlightened industrialist, who, in 1848, built the Phalanstery, which remained active until 1939. This project remained like an "on-scale" version of the Phalanstery with a very planimetric similarity, but with at least three differences to point out: the agricultural feature is lost in favor of a markedly industrial one; every family has a private accommodation; and the levels of community sharing are reduced.

3.3 Asia

3.3.1 Tulou

Opposite the island of Taiwan, on the Chinese mainland, is the fantastic province of Fujian, a mountainous region with the great coastal metropolises of Xiamen and Quanzhou. In the valleys of this province, a very characteristic (and extraordinary) building typology can be found, namely that of Tulou: residential buildings that are the physical representation (1) of the idea of community for the local society, historically based on clans, and (2) of the relation with the surrounding environment. In fact, regarding the first aspect, these buildings had a strong communitarian

3.3 Asia

Fig. 3.4 The group of Tulou Tianluokeng in Fujian (author 颐园新居; this file is licensed under the Creative Commons Attribution-Share Alike 4.0 International license; https://pms.wikipedia.org/wiki/Figura:Tianluokeng_Tulou_cluster_20140829.JPG)

value and defensive function, leading to "Tulou" being known as "a little familiar kingdom" or a "bustling small city." For the second aspect, as can be seen from all those pictures that show these buildings in their context, this typology always demonstrates an appropriate and authentic relationship with the natural environment (Fig. 3.4). In fact, it is very common to see these communitarian buildings, whose name "Tu lou" literally means "earth building," explaining the constructive technique used to realize them, located in Fujian's valleys, in a balanced equilibrium between the slopes, the river, and the sky. Maybe their contemporary appeal is really due to their design principles that follow the ideas of traditional art "feng shui,"[12] according to which the design of human settlements had to follow environmental equilibriums: a lesson of awareness unquestionably left behind by the contemporary practice. It is supposed that these astonishing residential fortresses were built between the twelfth and twentieth centuries, initially as residences to accommodate the members of the clans, the most privileged families of the society, containing a very great number of residents that, it is estimated, could in some cases reach 800 units. For some cases, since there is a large amount of well-stocked documentation and since the family clan often remained the same from the thirteenth until the twentieth century, we can even know the name of the founder of the buildings. Even with the changes in the Chinese society, the experience of Tulous did not stop with

---

[12] "Wind and water": a tradition that allowed man to find the right equilibrium with nature and with environmental forces.

the end of the Chinese empire: the construction of Tulous continued until the 1970s, for example with one of the 13 Tulous of Hekeng, built in the 1970s [10]. Despite this aspect, the isolation of Fujian Province, the emigration of the Hakka and the lack of infrastructure kept these buildings undisclosed until they were discovered by the international community between the 1950s and 1980s of the twentieth century—the same period in which, by the way, depopulation of the Tulous began because of the rural migration toward the cities. Today, some Tulous are abandoned, others occupied by a few people and others, those whose location allow them to be more easily reached, have become tourist attractions. Being so unexplored, the number of Tulous is still quite uncertain: National Geographic, in an article on these Chinese buildings, refers to an evaluation made by Huang Hanmin, one of the first researchers on this historical typology, according to which today there would be 2812 buildings, a thousand less than those previously estimated. Forty-six Tulous, built between the fifteenth and twentieth centuries, in the south-west of Fujian, have now been included in the UNESCO list.

Although Tulous were addressed mainly at the ruling class and the expression of political power, they cannot be considered icons of the richness and power of the families; on the contrary, all of them look almost the same and are constructed with very common materials. But notwithstanding this triviality, it must be understood that they are extraordinary examples of collective life, with private spaces, equal for all the members of the community, and a collective central space equipped to host the activities of the community life. Moreover, the majestic dimensions of the building and its compact walls represent very well the role that these buildings had in representing safe protection for the whole community and closure to a hostile context. In fact, as the image of a fortress indicates, their main function was definitely to defend and to guarantee the security of the clan, which had to protect itself from various and constant threats. For this reason, there is only one access and very few windows (more or less wide) starting from the second floor. For the same reason, the walls, which cover several floors, are made from a mixture of clay, limestone and sand that, on drying, becomes extremely hard, so as to resist several centuries of attacks, atmospheric agents, and earthquakes. Again, in the National Geographic report, a builder's son reports that the construction of a single floor could take up to one year of work. The buildings, which generally have a circular or square plan, were divided vertically between families. This means that to each family was assigned a vertical segment of the building, which started from the ground floor and reached up to the top floor. Each family could occupy one or more rooms, all identical, regardless of their role within the community.[13] Indicatively, on the ground floor there is an area with a kitchen and living spaces, on the first floor a warehouse and in the upper floors the bedrooms. All the space is maximized for living so that for each floor one external gallery, which the rooms face, is the only horizontal distribution system present in the building. Moreover, this walkway also functioned as a

---

[13] This is even more significant if we consider what the document produced by UNESCO reports: "The Tulous, although providing communal housing and reinforcing the structure of clans, were, until the twentieth century, mostly built and owned by one powerful individual" [10].

3.3 Asia

shared space for small storage. In the center of the courtyard, common activities took place and the space could be organized differently, depending on the Tulou: those without an open courtyard could host common buildings, or just a room for rituals, or shelters related to agricultural production. In any case, it was here that the activities of communitarian life took place, and for this reason, in contrast to the exterior, here the interior is extremely rich and full of symbolic references. Tulous represent close cultural relations among the members of a community and with the environment: the Tulou can be considered an example of human settlement in which traditional buildings combine forms of communitarian life, defensive organization, and an extraordinarily harmonious link with the environment.

3.3.2 Kibutz

Although in this chapter mainly historical cases are presented, where the term historical usually evokes a quite distant past, generally referring to a way of living previous to the industrial one, this case study is relatively recent but very interesting and is legitimated to appear here by several features that relate this experience to "past" cases: (1) the very interesting social structure and the role of the individual within the community; (2) the mix between political beliefs and religious common background; and (3) the relation with the territory (or even their role in defining the territory) and with agricultural production. In this first part of the description, unless otherwise stated, the description refers just to the first realities of the early twentieth century. In Hebrew, the word "kibbutz" literally means "communal settlement" and, in fact, they are communal settlements, born as rural, based on mutual aid and social justice. At least in the first experiences, their economic system is based on strong internal principles of property sharing, equality, and cooperation in production, consumption and education. For all these reasons, the kibbutz represents perfectly a society in which the involvement is based only on the free will of the individuals to participate in this community, which requires high levels of engagement, offering communitarian welfare and security feelings. The cooperative lifestyle, the decision-making, the management of the economic resources, and even the shared responsibility for the children's education make the kibbutz a unique experience in the panorama of community living. Degania (Fig. 3.5), the first kibbutz, was founded on October 28, 1910 by ten men and ten women, who "founded an independent settlement of Jewish workers. A cooperative, without exploiters and not exploited: 'a commune'".[14] These people, with Jewish roots, came from eastern Europe with the dual aim of finding a link with the original land of their ancestors and establishing a settlement arranged on an equitable way of life. Since then, 38 years before the foundation of the state of Israel, to date 273 kibbutizm have been built (half of them before 1948) and members of the Zionist Youth Movements, coming from all over

---

[14] Sculpted words to celebrate the establishment of the kibbutz "Degania," in the north of Israel.

Fig. 3.5 General view of kibbutz Degania Alef with the sea of Galilee. (Author Zoltan Kluger (1896–1977); taken on 01/10/1946; available from National Photo Collection of Israel, Photography Department Government Press Office, under the digital ID D516-056; This image is now in the public domain because its term of copyright has expired in Israel. According to Israel's copyright statute from 2007, if the copyrights are owned by the State, not acquired from a private person, and there is no special agreement between the State and the author, a work is released to the public domain on 1 January of the 51st year after the creation of the work (paragraphs 36 and 42 in the 2007 statute); https://commons.wikimedia.org/wiki/File:GENERAL_VIEW_OF_KIBBUTZ_ DEGANIA_A_WITH_THE_SEA_OF_GALILEE._ה_ללכ_לש_יוביק_דגנה_א%27תרגולת_הכנרת. D516-056.jpg)

the world, founded the majority of them. In fact, as evidenced by Adam Pavin [11], the Zionist component played a major role in the development of these realities, due to their policy to define the borders of Israel with rural settlements, even where the settlement would have been in a total desert area. The "perimeterial" position in relation to the center of the state of Israel[15] led the founders into a hostile natural environment. In general, a common design pattern to describe how the majority of the kibbutz settlements are organized could be recognized: the residential buildings surround the common spaces and facilities, which can be easily reached by walking or cycling. These common facilities are, usually, the children's houses, a playground for every age group, a dining hall, an auditorium, a library, a swimming pool, a tennis court, a medical clinic, a laundry, and a grocery. Outside this residential area are

---

[15] Eighty percent of the population lives in the north and south of Israel [11].

all the production activities, the agricultural fields and the fish ponds. The production activities of kibbutzim are organized in different branches, and now they have been expanded to industrial sectors or rural tourism (thanks to the great naturalistic environment in which the kibbutzim were settled). Usually, following a methodology of consensus decision-making, a general assembly manages the business of the kibbutz (from policy and economic affairs to the approval of new members), while the everyday management is supervised by elected committees. In this panorama of total equality, three aspects are relevant: (1) the position of women, which was identical to that of men; (2) the growth of the children, who lived together in the common houses and their progress were followed directly by the whole community (nowadays, they still spend most of the time in the communitarian structures with their peers, but they grow in the family's house); (3) the economic aspect related mainly to personal incomes.

Actually, the strong changes in the contemporary society are also causing changes in the social and physical structures of the kibbutzim that, in trying to survive, must adapt by assuming different shapes. For example, some years ago, in the first established kibbutz, 85% of the residents voted to abolish the collective organization. In one article, Davide Frattini [12] emphasizes that the economic crisis of the kibbutz turned into a recession of ideals.

According to the Kibbutz Movement, the way of distributing the income among the members of the kibbutz is the most important difference among today's kibbutzim, thereby creating different configurations. (1) Communal Method (60 kibbutzim), where the sharing of income is equal among the members; (2) Integrated Method (20 kibbutzim), where the income of each member is composed of three parts, whose proportions can vary from case to case, but are divided like this: one equal for everyone, one based on the member's seniority, one based on the individual contribution to the community; (3) Security Net Method (190 kibbutzim), where the personal income is mainly based on the private salary, while a percentage is dedicated to the kibbutz management and to economically support those members with an income lower than the minimum value defined by each community.

After the crisis of 1985 that caused a departure from the world of the kibbutzim, a gradual return to growth in the number of residents has been recorded (Table 3.1). In 2004, 2.1% of Israel's population lived in a kibbutz, making a total of 116,000 people in 266 kibbutzim. From 2004 the data indicate an aging population (30 years on average, compared with 25.8 years in 1989) [11]. Also, the fact that the kibbutzim have always played a fundamental role in the history of the state of Israeli is relevant: from the definition of the boundaries to the absorption of new immigrants and to service in the armed forces. The kibbutzim also had a very strong influence in the political life of Israel: in 1949, in the first Knesset, 26 members of the Parliament came from the kibbutzim, but in the elections of 2013, just one kibbutznik became a member of the Parliament.

This experience of a communitarian way of living, pushed to the extreme to share within the community even the growth of the children, is very interesting. It represents, maybe, one of the last strong attempts to recreate communities, where

74   3 Sharing as Cultural Preexistence

Table 3.1 Number and population of kibbutzim in Israel since 1910

Year	No. of kibbutzim	Kibbutz population
1910	1	
1920	12	805
1930	29	3900
1940	82	26,550
1950	214	67,550
1960	229	77,950
1970	229	85,100
1980	255	111,200
1990	270	125,100
2000	268	117,300

the individual can find their own realization far from the individualistic modern world: from the rise of a dream in the early twentieth century to the cooling down of the main values in the contemporary years.

3.4 Africa

3.4.1 Matmata

Although the aspect of sharing in this project is not as explicit as it is in other projects presented here, it seemed correct to present it in this review because it represents the result of an adaptation (1) to a difficult natural environment, and (2) to a social environment, such as the Arab one, characterized by specific social needs. As Paul Oliver notes [13], speaking of the cultural aspects of vernacular architecture, "[a]s has been noted, this was a conservative community with strong moral codes related to Islamic law. The protection of women from other men, comparative seclusion of women and general family privacy were valued highly." Hence the need to clearly define the social bounds of the "family" and the physical limits of the house. Nevertheless, this settlement generates some interesting issues of community life.

The geographical area of Matmata Plateau is situated in Tunisia, in northern Africa, about 20 km west of the Gulf of Gabes and borders with Djebil National Park. It is an area very poor in rainfall but which, due to the characteristics of its topography, allows excellent conditions for defense. This area has historically been inhabited by the nomadic and semi-nomadic population of the Berbers, which has gradually become a more permanent population, although it partially conserves pastoralist economies and has never completely given itself to trade, unlike the nearby urban communities of Gabes, Medenine, and Zarzis. In this area, two types of underground constructions are settled: one with a vertical development, with a central patio, about 10 m deep, in which the houses only have a view toward the patio; and another, with a horizontal development, located on slopes, with a direct view to

3.4 Africa

Fig. 3.6 Ancient Berber architecture in Matmata (Tunisia) (author Sarah Murray; taken on 11/06/2011; licensed under the Creative Commons Attribution-Share Alike 2.0 Generic license; https://commons.wikimedia.org/wiki/File:Tunisia_matmata_pacio.jpg)

the outside (Fig. 3.6). As highlighted by Gideon [14], these Tunisian subterranean communities show very well the human ability to adjust to the environment: a hard climate, the quality of the soil, and a need for military defense contributed to creating these settlements. The problem of defense against foreign invasions has in fact greatly influenced the Berbers' way of life. At the time of the first violent contacts with the Romans, these people took refuge in the most inaccessible areas of the region, going first to occupy the natural caves and later living in caves artificially excavated. At a later stage [14], these populations began to settle in less inaccessible areas, although still mountainous, where they placed fortified centers for the protection, in particular, of granaries. These two forms of settlements began to evolve simultaneously and became more and more common for several centuries. The third phase, the one we are interested in, is the development of underground dwellings that began to spread in increasingly less mountainous, if not flat, areas as the Berbers could develop more friendly relations with invaders, in particular with the Arabs. This building typology, which is the one present even today, started with the Berebers and evolved for about 13 centuries, mixing with the Arab culture. The following description will deal with the vertical underground solution, which can be clearly observed in Matmata. Here the Berber-Arab community developed its living space, seeking protection from a hostile natural environment and responding to the demands of a social environment.

Although, as previously said, in a culture like the Arab one, the margins of sharing spaces are limited for well-defined cultural reasons, the spaces of Matmata assume a strong sense of identity and community. Hill and Woodland [15], for example, report that this type of construction has entered so much into the sense of belonging to a community that, when in Chenini (a similar complex in Matmata) new modern dwellings, built by the Tunisian government to provide better housing facilities for residents, were inaugurated, people decided to abandon them and to return to living in traditional structures.

According to Gideon [14], in 1972 there were 600 households in Matmata, of which only 60 were uninhabited. Matmata community can be divided into four districts, each of which has its own characteristics depending on the main clan to which it refers. In fact, pit dwellings are dispersed in the landscape but organized in neighborhoods according to clans; nevertheless, houses develop without a precise planning principle. The dwellings can have a quite significant extension, covering an area of about 30 × 30 m, to which is added another 40 linear meters of distance between one well and another. The distribution is influenced by the Arab and Middle East building tradition: a central patio surrounded by rooms, allowing an excellent level of privacy, particularly for women who, in Arab culture, can unveil their head only in the privacy of their home. In fact, the patio plays a central role in the life of the family community, since the Islamic principles are imposed to build privacy into the family home [16].

The rooms, with an ovoid or rectangular shape, develop radially around the patio and their number depends on the size of the family and the additional needs to be met (providing protection for animals, storing grain, protecting utensils, etc.). The largest room is that of the head of the family and is usually located 1 m above the surface of the patio and looks to the east, in the direction of Mecca. The patio is not only the central place that the rooms overlook, but first of all it is the place where families can carry out common activities and, at its center is usually placed the cistern that stores rainwater. When the community expands with the marriage of a child, the spaces expand with the construction of a new patio that connects with the original one, thereby creating a system of shared spaces belonging to the same extended family. In fact, extended families used to occupy single rooms and houses were extended as families grew, so that the size of the houses depended on the clan's structure.

In addition to the "privacy factor," the fact that they are dug deep into the ground makes it possible for these dwellings to have an excellent level of thermal insulation, to minimize solar impact, and to protect from wind storms. Furthermore, these buildings require very little water for construction and very little maintenance, making this construction practice extremely sustainable.

In essence, three community levels can be defined for this type of settlement. (1) The lowest one is the extended-family community level, which goes beyond the Western concept of the nuclear family. However, this concept, gradually introduced also in this society, increasingly undermined the lifestyle of these communities. This community level finds its spatial significance in the patio, which represents shared life. (2) The second community level is that of sharing public facilities. Beyond the threshold of the houses extends the public space, whose space belongs to everybody, except for an area around the wells that is culturally accepted as semi-private. In this vast shared space, the community develops small forms of agriculture and all the common structures belonging to all the members of the community are located here: a protected storehouse, shared by several families (*ksar*), the market and the well. The communities of Matmata evolved with a very close relationship with this kind of shared environment, creating a strong sense of collaboration between the community's members. For centuries, for example, the only active well was the one of the community, located in the lower part of the hydric basin. Furthermore, based on the field interviews conducted by Hill and Woodland [15], both in Matmata and Chenini, communities interacted intensively to cultivate fields, sharing resources and risks among the individuals. (3) Finally, the third community level is that of the cultural and religious landscape, since these constructions define a landscape with a strong cultural connotation. Before the landscape was blemished by recent buildings outside the ground, the landscape was a combination of the aridity and the artificial craters of the patios that dispersed into the landscape leaving one to guess the boundaries of the village. Here the only structure that stood out in the landscape was the cultural and religious symbol of the community: the mosque, with its minaret standing out against the landscape and uniting the community, through an organization of space and time, with the Muezzins, who articulated the phases of the day and called the faithful to pray. In conclusion, as Hill and Woodland [15] argue, for the inhabitants of these subterranean villages, the houses were a symbol of "family unity, effort and history," since they represented the well-being and the security attained within the social and physical dimensions.

3.4.2 *Umuzi or Kraal*

A second typology of historical settlement that is present in many populations of Africa (particularly in the sub-Saharan ones) and which assumes a strong social significance is the "village." Although these types of settlement are multiple and vary greatly depending on the geographical area and the populations they belong to, there are some common features that will be highlighted through the presentation of the case of the Zulu people.

The Zulu people are an ethnic group from East Africa with a well-devised social structure, which is based on extremely precise and strict rules of conduct that condition the lives of all members of the community, from the children to the leaders. The Zulu society is structured in extended families to which other smaller nucleuses are joined, sometimes even forced by social duties. The Zulu settlements define very well the sense of belonging of an individual to a community, if not even the very sense of existence of the individual. In this southern area of the continent, in fact, indigenous settlements in general have a great functional, organizational, and symbolic richness. In this regard, Christopher Ejizu highlights [17] two very important aspects to be emphasized in this discussion concerning the influence of communities on the individual: (1) People who have left their original villages and communities need to find relational constraints and thus seek to rebuild these ties through affiliations to clans or to ethnic or religious groups; (2) For many African cultures, the concept of an individual outside a community is hardly understandable: the individual exists only because they are part of a community. Also, in Umali, we can read, as in the African cultures, that the community is the custodian of the individual [18]. This concept of the individual is extremely relevant because it explains how humanity is connected both to the physical and the incorporeal environment, through the awareness of belonging to a group, composed of the people who live with them, the ancestors and the souls of those who still have to be born. The bond with the ancestors is indeed very strong and is lived through numerous rites that mark the time and the spaces of the community, where a significant space is always dedicated to this mystical encounter between the spiritual world and the earthly one. Around this space in which the community, connecting with the spiritual dimension, meets its realization, the other common activities of the group also take place. The village, therefore, takes on a very significant meaning for the African individual, being able to culturally evoke human relationships, mutual support, the sacredness and meaning of existence, and a sense of time.

The organization of Zulu villages is based on a circular planimetric design. As evidenced by Laburn-Peart, the circular shape was one of the remarkable physical characteristics of most South Africa precolonial settlements. This shape was often associated with the cattle kraal or enclosure, which was surrounded by the clan's huts, so the term *kraal* could refer to the village itself [19].

Specifically, kraal refers to a tendentially circular enclosure used by the Zulu populations, primarily to protect livestock and, secondly, to become the system on which the village was based. In fact, the village tends to consist of two concentric rings of palisades (Fig. 3.7). The inner circle defines the space for the protection of livestock, while the outer ring is the one occupied by the community's homes. These

Fig. 3.7 A Zulu kraal in South Africa (unknown author; credit licensed under; this work is a faithful photographic reproduction of a two-dimensional, public domain work of art; it is in the public domain in its country of origin and other countries and areas where the copyright term is the author's life plus 70 years or fewer; https://commons.wikimedia.org/wiki/File:Kaffir_kraal.jpg)

villages are usually positioned on a slight slope that allows rainwater to clean the soil from animals' excrement and move it away through the entrance door, located in the lower part of the slope. Linked to the images of the lance and the shield, the figurative aspect of power, related to the left and to the right, is also highlighted as symbols of strength and protection (one attacked with the right and defended with the left). In the Zulus' villages, the dwellings (iQukwane) are placed in accordance with the inhabitants: the mother of the head of the clan resides in the largest hut, which also has the function of altar of the family's ancestors, and is placed on the side opposite the entrance; the hut of the community leader is on its right side, and the wives' huts on the left and right sides according to the importance of the marriages. Near the entrance there are the huts for the youngest, not yet married, with a distinction between the huts on the right, inhabited by the males, and those on the left, inhabited by the females. There are also some constructions that play the role of granaries and watchtowers. The community ties within the village are so strong that even if everything is based around the leading figure of the clan leader, the children grow up addressing all the adult and married members of the community as if they were their own parents, to the point of calling them "mom" and "dad."

3.5 America

3.5.1 Calpulli

Calpulli is a Nahua term, which literally means "big house" or "group of houses," and refers to a social and spatial organization of the Central American society under the Aztec Empire, which covered from the fourteenth to the sixteenth centuries. The term has been the subject of numerous debates among the anthropologists of the last century, which were divided on the reading that should be done on the *calpulli*: those who sustained that this was a mere territorial organization (as believed by White (1940) and Moreno (1931)) [20] and those who thought that the *calpulli* was also a social organization (Bandelier (1878), Thompson (1933), Vaillant (1941), Murdock (1934), and Caso (1942)) [20].

To deal with this topic, reference will be made to both the social and architectural organization that this concept refers to. It must be said that what is known about the political and social organization of the Aztec civilization comes mainly from the chronicles of the Europeans, which, as regards the explanation of the *calpulli*, are particularly lacking. Furthermore, a *calpulli* is a structure belonging to different societies/ethnic groups belonging to the Aztec Empire and was developed over a fairly long historical period [21]. The uncertainty given by the limited amount of news and this situation of variety [22] makes it difficult to be very precise and detailed in the description of the *calpulli*, but three basic elements that concern the research remain clear: (1) the central role of this community organization within the society and territory of the Aztec period; (2) the fact that this organization referred to an understanding and organization of the world by the Aztecs; and (3) that each individual had a strong sense of belonging to this social unit. Because it is so significant and basic for the Aztec culture, the *calpulli* is a concept that can refer to different contexts: it can be associated with a rural and urban community [23]. In fact, as previously mentioned, the *calpulli* is not only a physical structure for spatial organization, but it is the most elementary unit of social organization that allows control of the territory, be it rural or urban. The control of the territory in the Aztec Empire is something similar to a European feudal structure, even if it seems that it was characterized by a more complex bureaucratic system. To better understand the role of the *calpulli* in the complicated Aztec society, it would be helpful to briefly introduce the levels of territorial management. (1) At the lower level, there were rural villages scattered throughout the territory and managed by a village chief or a council of elders. These social units assumed the organization of a *calpulli*. (2) At a second level, the city states (*Altepetl*) were under the control and influence of one of the three capitals. These city states were run by a dynastic power and had influence over the surrounding area. They were also organized under a system of *calpultin* (Nahua plural of *calpulli*). (3) At the highest level, there was a confederation of the three most important cities of the empire (*Huetlatoani*): Tenochtitlan, Texcoco, and Tlacopan. Of these three, in the last century of the empire, Tenochtitlan was the city with the greatest power, becoming the central city of the Empire, which fell when Tenochtitlan capitulated (1521) [24]. The neighborhoods of these capitals were organized into smaller elements that were the *calpultin*. In the case of this capital, for example, the city was divided into four main areas, each of which was organized in *calpultin*.

The role as the ultimate element of social organization suggests the importance the *calpulli* had in conceiving social relations within the empire. The relations were based on interactions between peers that supported each other in daily activities. Therefore, whether in rural or urban areas, the conception of the role of the individual in the social environment did not change: in both cases, the individual was part of a limited nucleus of people with the duty to contribute, in terms of labor and economic support, to the wealth of the *calpulli* and indirectly to that of the wider society, up to the empire. Despite the common role of the *calpulli* in standardizing the position of individual and community within the Aztec society, differences in the composition of the *calpultin* should be pointed out. It is known, in fact, that these settlements could be composed of the members of the same family, of the members of the same clan, of the members of the same ethnic group (being an empire, this included different ethnic groups), or even of the citizen members connected by the same job, as they were guilds.

It is interesting how this unit functioned independently and communally: each *calpulli* had to ensure its maintenance and could count on productive elements shared among the various members of the community, which guaranteed its self-sufficiency. In particular, it was common practice to share agricultural fields or *chilampas* (floating cultivated structures, particularly famous in the city of Tenochtitlan, which was located in the middle of the large Texcoco lake). *Calpultin* were organized through a series of connected patios that linked the community members'

3.5 America

houses; as the community grew, patios and dwellings were added. The dimensions, therefore, were not always fixed: as demonstrated by Michael Smith [25] regarding the city of Molotlan, the *calpulli* was composed of 128 households, divided into 9 wards ranging from 1 to 32 households in each. Each *calpulli*, however, could count on its own administrative center, a religious complex, a market and barracks, where young people were trained for war. So for the whole society, the belonging to the empire passed through the belonging to the *calpulli*, which was the structure closest to the individual and the structure in which they recognized themself. Noting the importance of belonging to their own *calpulli* is to note how the warriors coming from the same *calpulli* fought in the same military unit. Little is known about the administration of the *calpulli*, which were managed by a leader elected by the community or descended from their predecessors, supported by a college of elders. However, those who managed the *calpulli* (*calpullec*) were exempted from agricultural or productive work so as to be able to dedicate themselves totally to the management of the community.

After the fall of the Aztec Empire, *calpultin* continued to be a fairly strong form of community organization and, even in the colonial *aciendas* of central Mexico, some legacies of the *calpulli* can be found. In essence, the concept of the *calpulli* not only represents the Aztec political and social organization but transcends toward an aspect of relationship with the environment, in which the individual, through the community in which they live, becomes part of an alliance with the environment. The charm of the *calpulli* lies precisely in the beauty of its understanding of the world, in which the basic element is the individual who is part of a community of individuals, capable of managing the surrounding territory, be it urban or rural. The *calpulli* represents a way of understanding the world that takes shape in social and urban structures able to provide approval for the individual and the environment.

3.5.2 Raramuri Settlements

Raramuri are one of the most numerous populations among the North American natives. These people have been pushed by the advance of the European colonizers up to the mountainous regions of the western Sierra Madre, part of the Pacific coastal chain, which extends between the current states of Arizona, Sonora, Chihuahua, and Durango. Within this region, Raramuri are concentrated mainly in the state of Chihuahua. This population, which is commonly known firstly by the conquerors and now by the Mexicans as "Tarahumara," has a current estimated population of 50,000/75,000 individuals and is famous for its athletic skills. In fact, the world "Raramuri," as they call themselves in the indigenous language, means "speed foot" or "good runner," referring to this exceptional capacity for physical endurance. Raramuri are able to face extremely long runs, to reach isolated settlements or even to chase large prey that after tens, or hundreds, of kilometers fall victim of the Raramuri due to physical exhaustion.

Raramuri society is based on a very close relationship with the natural environment; in the beginning this was because their religion itself was very much tied to the environment and to the attention toward the natural world. Since coming into contact with the Christian religion and the numerous conversions that have taken place, attachment to the environment has been preserved through a form of life and a social attitude that is absolutely respectful of environmental resources [26]. The agricultural production is in fact quite limited and essential (corn and potatoes), as is the exploitation of natural resources. Just as Raramuri individuals bring great respect to the natural environment, so they also bring as much respect to the social environment: they are a very shy and taciturn people and it is difficult to imagine them raising their voices. The Raramuri society is organized for families that work very closely in these agricultural productions, making the practice of sharing activities and resources the basis of social coexistence. In this regard, their language is very significant, because rich in some respects conveys the Raramuri way of understanding the world and society. For example, the word "teaching" is expressed with the concept of "sharing knowledge," or when they beg (as unfortunately happens in cities where the Raramuri seek fortune too often without success) they ask to "share." This is because the background idea of life, for Raramuri, is sharing resources: both to lean on each other in a hostile territory and because it is difficult to think of the possibility of possessing more than the necessary. According to the life philosophy of the Raramuri, in fact, it is difficult to conceive the concept of storing, because you take from the environment what you need and nothing more. For this reason, in the quest, for example, that money (or goods, foods, etc.) which is exceeding to a person and that will not be used can be given to "charities," in a simple gesture of "sharing." While this form of understanding the use of resources makes Raramuri society extremely sustainable, their resulting way to understand the concepts of "storage" and "exciding" put their existence at risk several times because they often found themselves in the situation of not having the resources to face particularly harsh winters or particularly dry seasons. This form of conceiving life and social relations, on the one hand, comes from a millenary-old way of seeing the world, of understanding its phenomena and adapting to them, and on the other, it marks the way in which they organize society and the territory in which they live [27].

Social organization is based on communities and extended families, each of which occupies a vast territory in the valleys of Sierra Tarahumara. Within these communities, the smaller elements are the nuclear families, comprising a few people who, still today, live in caves or in very simple huts, made of foil, tree trunks, or concrete blocks. These houses can be set up to hundreds or thousands of meters away from each other (Fig. 3.8). Between one house and another, there is the territory that is partly cultivated by families in a shared way: the member of the community moves to cultivate the land of each of the members, supporting one another, and consuming from time to time what is harvested based on the season. There are two aspects that should be highlighted: (1) the tendency to share possessed things, be they physical or ephemeral, such as culture or work; and (2) the tendency not to

3.5 America

Fig. 3.8 Raramuri settlement in Barranca del Cobre, in the Sierra Tarahumara (Mexico) (author Emanuele Giorgi)

exploit the territory for production, but to take from the environment only what is necessary. This means working the land only for what is necessary to sustain the community.

The case of the Raramuri is presented here because of this fantastic way of understanding the relation with the environment and the way of developing social relations. At the same time, it is interesting to understand how, parallel to a very strong sense of sharing and mutual support, the housing settlements are divided into single units. Each individual has their own space, each family their home, and the housing units are a long way from each other. Therefore, there is a way of developing settlements that is totally different from those that this review proposes, in which equally strong feelings of sharing materialize in strongly shared spaces. The need to live in a strong relation with the natural environment and the need for spatial individuality bring people to look for insulation, but with the strong consciousness of belonging to a wider and indispensable community. The sense of community is perceived so strongly that, if one of the members leaves the rural community, looking for opportunities in the cities, it is very difficult for the community to easily accept their return.

3.5.3 Shabono

The Yanomami are one of the most isolated populations of the contemporary world, lost in the Amazonian rainforest over an area that extends between northern Brazil and southern Venezuela, with an estimated population of almost 30,000 individuals. They are known to be incredible botanists, experts in the thousands of varieties of plants that are the major source of livelihood of this people: a source of food, but also of utensils, building material, poison, perfume, etc. It should be noted that the Yanomami's use of forest resources is limited to the needs they have: knowing that the forest needs time to regenerate, they use what is strictly necessary to satisfy the needs of the community. According to Milliken [28], this is one of the biggest differences with contemporary culture: as they know the environment in great depth, so they know how to manage it. While the Yanomami are a fairly aggressive people, particularly toward the outside of the community, the management of social relationships is based on an incredibly strong equity between people. There is no figure of a recognized leader and decisions are taken through internal debate, with a practice similar to the one we know as "consensus." Furthermore, the idea of sharing resources between the members of the community and managing shared activities together is very strong. For example, in addition to community hunting, fishing trips, and communal cultivation, there is the custom that a hunter cannot eat the prey they kill but must share it with other members of the community, receiving, in return, food from another hunter. This practice certainly contributes to establishing a strong sense of community.

In the case of the Yanomami populations, the physical space that houses the community is represented by a single large structure that fits into the landscape of the Amazon forest with an amazing compositional beauty (Fig. 3.9). The simplicity of such a pure design, combined with the light and dark colors of the dry palm leaves, contrasts with the unique elegance of the majesty of the forest. This structure is composed of one or two sloping roofs that basically rest on the ground on one side, and on a wooden palisade on the other. This timber structure, covered by palm leaves, develops in a circle around a large central space, used for common events, parties or rituals. This ring, with an indicative width of about 10 m, provides shelter for numerous families, each of which occupies a sector of the structure that can accommodate, depending on the circle's diameter, from 50 to 400 people. Generally, the community is physically closed to the outside, or by the external pitch that reaches the ground or by a small wall that closes the vertical space between the ground and the external pitch. Clearly, this perimeter closure is not total, as some openings (usually four) are left to allow the connection between the central space and the surrounding area, which is usually cultivated primitively by women. The ring covered by the structure is essentially divided into two areas: an inner ring used as a common area and a covered "distribution system," and an outer ring, which is the most private part of each household. In this area of "belonging" to a family, each nucleus runs its own cooking fire, around which family belongings are arranged: some hammocks on which to sleep and a few other objects. However, the relatively

3.6 Oceania

Fig. 3.9 Aerial view of a Shabono in the Amazon forest (Roraima, Venezuela) (author Marcos Wesley, CCPY, 2005; licensed under the Creative Commons Attribution-Share Alike 4.0; International license; https://commons.wikimedia.org/wiki/File:Yanonami_-_Amazonas.jpg)

simple structure has some not-so-trivial constructive aspects. For example, the surface of the covered living area, in rammed earth, is located at a level slightly higher than that of the central esplanade, thereby allowing a drier and healthier residential space. *Shabonos* are built recurrently every few years, adjusting their size according to their population, and sometimes multiple *shabonos* are built close to each other, forming larger communities.

3.6 Oceania

3.6.1 *Marae*

Maoris are the main indigenous population of New Zealand, characterized by a culture that influenced the Western conquerors more than happened in many other parts of the world. Characteristic of this people is their conception of the environment: totally different from the Western one and, in general, difficult to understand for all of those who do not belong to the Maori culture. According to Merata Kawharu [29], the term "environment" can be translated as "*taiao*," which in turn would mean, according to the Williams Māori dictionary, "world, country, district," thereby relating the term "environment" to economic, political, cultural, and spiritual dimensions. Moreover, in the Maori conception of the environment, there is no clear distinction between the temporal and the spatial dimensions, which join together in a twine of awareness of the use of environmental resources and of responsibility to follow the footprints of the ancestors [30]. These complex relationships that govern the way of conceiving the world are absolutely fascinating. Here there is no opportunity to go beyond these few lines of introduction, but it is a theme that can be deepened thanks to a very rich (albeit complex) bibliography.

In this society so devoted to the relationship with environment and time, the architectural complex of the Marae takes on an extremely important role: the center of the community that has slightly evolved over time, while always maintaining its central role in the lives of individuals, where the community merges with the environment, the history, and the ancestors. Although so central to the life of the Maori communities, little has historically been written about the role of the Marae in the Maori society—partly because we do not have any written documents on this subject prior to the arrival of the Europeans, and partly because the first explorers did not report any details about this. Even scientific production has been scarce on the subject [30]—up until the 1950s when more and more attention was directed toward these buildings whose symbolic and cultural richness fascinated observers. "Marae" is a term that in Polynesian cultures generally refers to a sacred place for pre-Christian societies, defined by a plain of stones bounded by a small perimeter wall and with some symbolic monoliths. Today, these places, which are considered abandoned, are visible on many islands in the Polynesian archipelago. In Maori culture, however, this same term takes on a much richer meaning and is still very much alive in the management of contemporary communities. While preserving the meaning of a delimited and sacred space (an outdoor place for meetings and Maori rites), it also becomes a community element, a symbol and extension of the environment and the landscape.

In essence, this architectural complex consists of: (1) a fenced esplanade (*marae ātea*); (2) a main community building (*wharenui*); and finally (3) other community buildings supporting the shared activities that the community carries out in the Marae, such as a kitchen, a dining room, showers, toilets, etc. Austin [30] clearly describes how these architectural complexes are positioned in the territory according to strong rules: the esplanade, in fact, must be directed, on one side, toward an "open" landscape, which can be the sea or a valley, and on the other side, it must give back to a landscape that delimits, such as a mountain, a slope or a forest. Moreover, the Marae tends to be arranged in the direction of the flow of a river. On the side of the delimiting landscape, the community building is constructed, while other buildings can be added along the other sides, to better define the perimeter of the esplanade. The entrance to the Marae is always on the side of the open landscape, so that to proceed toward the community building, the esplanade must be crossed. Even from this very synthesized presentation, one can imagine how this architectural/landscape system is extremely rich in meanings and in relations with the environment that can be better understood only by comprehending the Maori culture in depth.

3.6 Oceania

Fig. 3.10 Men performing a haka on Te Papaiouru Marae, Ohinemutu, Rotorua, in about 1908. (Author Thomas Pringle; this work is in the public domain in its country of origin and other countries and areas where the copyright term is the author's life plus 70 years or fewer; https://en.m.wikipedia.org/wiki/File:Haka1908.jpg)

Historically, the Marae was inhabited by the whole community, as it was "one big family" we could say, with a Western vision: the main building was the home for the entire group, where the members of the same community shared spaces and activities. Now, with the diffusion into New Zealand of the concept of the nuclear family, their function has changed: from a common home, where the whole community slept and ate together, the Marae is nowadays a symbolic place for the community, where the most significant activities take place and also where people can still sleep and eat together, but not daily. Being the "home" and the "physical and identity heart" of the community, the Marae must be protected and the access to the common building is guarded very jealously. The welcome ceremonies (pōwhiri) that take place in the esplanade, dedicated to the god of war, are famous (Fig. 3.10). During this event, the community dances and sings, while a warrior from the host community "challenges" the guests to see if their intentions are peaceful, or hostile [31].

If the guests arrive with peaceful intentions they can proceed to the heart of the community, the common building, entering through the porch, which has many symbolic meanings. Being the threshold that separates the place of the god of war (the esplanade) from the place of the god of peace and agriculture (the community building), the porch endows the strong symbolic tension to cross from a violent space to a peaceful one [32]. The whole building symbolizes a human body (sometimes a specific ancestor): in the facade, the pinnacle (*tekoteko*), the dispro-portionate roof beams (*maihi*), and the side columns (hook), respectively, symbolize the head, the arms, and the legs of the human figure. The interior of the common building is substantially free from furnishings, but rich in ornaments and engravings that describe the history and genealogy (*whakapapa*) of the community. The central column (*poutokomanawa*), whose structural function to support the weight of the roof is not always relevant, works mainly in symbolizing the heart of the ancestor, and more generally, of the community.

3.7 Conclusions

Although these cases are not absolutely exhaustive for all the typologies of com-munitarian settlements that can be found in the history of human societies, these examples allow some reflections to be generated about shared living and its influence on the spatial needs and on the relations between individuals and the envi-ronment. Comparing the examples, some aspects that associate one case with, or differentiate it from, the others can emerge. Among the differences, there are at least two aspects that can give rise to very short reflections. The first concerns the rela-tionships among the members of a community. It is not a banality to note cases where: (a) communities crystallize around very solid families and the relationships do not expand much further than the clan levels (Tulou, Matmata, Kraal, Raramuri); (b) the relationships between "family heads" are those that generate wider social relations within the community (Greeks); (c) the concept of the nuclear family becomes weaker to allow a strengthening in the relations between the community members (Cascina, Shabono); (d) community relations find their origin mainly in the economic or political structuring of larger organizations (Calpulli).

The second aspect concerns the physical dimension in which the community expresses its relationship with the environment and finds its completeness. From the study of the previous cases, these situations can be extrapolated: (a) the spaces of the community, both private and common ones, are not strictly connected, but com-munity relationships take place in a landscape or territorial dimension (Kibutz, Matmata, Raramuri); (b) the community must be physically delimited by a wall, a fence or a building, which isolates it from the external context (Marae, Calpulli, Kraal, Matmata, Tulou, Greci); (c) people find their expression thanks to an element that encloses the community, filtering (without isolating) internal and external spaces, and in particular allowing the required spatial continuity for agricultural activities (farmstead, *falansterio*, *shabono*).

Despite these two differences, there is a main aspect that highlights a commonal-ity among all the cases that must be highlighted to make a quick reflection: the purpose of isolating. We have just seen how these cases can promote different social relationships or adopt different spatial solutions. However, all the cases have a cohe-sion on the main reason that led to the generation of the community, which is the need to defend itself from an external environment. Both for the philosophical and the cosmological interpretation of the role of man in the world, for adapting to violent migratory flows, to protect individuals and livestock from the attacks of wild nature, it is precisely the need to defend itself that leads to the creation of protected communitarian spaces. This need promotes the construction of walls, fences, patios, compact buildings or, in the case of the Raramuri, taking refuge in inaccessible areas, where it is precisely the high inaccessibility of the environment that defines the community's space. And it is in this protected place that individuals can count on mutual support from other members of the community; this protection toward the outside promotes and reinforces the social relations toward the inside of the community. Traditional communities, therefore, can be understood as groups of people (linked by more or less strong family ties) who seek protection in a limited physical space in which to create relationships of mutual support. This is quite interesting if we consider that, today, the widespread idea of community is associated, in contrast, with the idea of openness and wide sharing, where the interactions do not end in the perimeter of the community, but tend to expand to the next context and beyond. For example, one of the characteristics of co-housing that we will analyze in the next chapter is indeed this aspiration to be promoters of new social relationships in the context where they are inserted. In the same way, other forms of contemporary communities are born out of the ambition to generate relationships between individuals but do not find their first reason in closing themselves off from the physical or social context. Indeed, today, when groups of people try to define a "community" by creating barriers to defend themselves from the surrounding context, they do not get a community as a result, but rather environments in which relations tend to be more limited than elsewhere. An example of this case is the gated communities, a phenomenon that increasingly characterizes the urban territories of all the continents. There is a lot to investigate and say concerning these gated communities. Firstly, to understand how these settlements called "gated communities" disavow the profound idea of community. Secondly, because it would be interesting to reflect on how those who are involved in public policy, territorial development and planning should make an effort to reinterpret these urban settlements, since nowadays, gated communities are considered serious problems for urban sustainability, but at the same time, it is hardly believable that markets move away from proposing a housing solution like this one that is so demanded and profitable. Finding a form of territorial and architectural design capable of uniting market needs with environmental needs should become a priority in the contemporary debate.

References

1. G. De Carlo, Sulla progettazione partecipata, in *Avventure urbane. Progettare la città con gli abitanti*, ed. By M. Scavi (Elèuthera, Milan, 2002), pp. 244–248
2. L. Trabattoni, L'abitare incerto. Gli spazi collettivi all'interno delle residenze ad alta densità. Doctoral Thesis. University of Pavia (2009)

3. M. Le Bel, Natural law in the Greek period, in *Natural Law Institute Proceedings*, ed. By M. Le Bel, E. Levy, G.H. Gerould, H.A. Rommen, R.N. Wilkin, vol. 2 (Natural Law Institute Proceedings, Notre Dame, 1949)
4. G. Reale, *A History of Ancient Philosophy, III, Systems of the Hellenist Age* (State University New York U.P, New York, 1985)
5. E. Greco, *La città greca antica: istituzioni, società e forme urbane* (Donzelli Editore, Roma, 1999), p. 9
6. M. Romano, Come progettavano i piani regolatori nella città greca e romana. Resource document. Estetica della città (2011). Retrieved from http://www.esteticadellacitta.it/pdf/cittaantiche.pdf. Accessed 10 May 2019
7. S. Crotti, G. Bertelli, M. Reggio, D. Vanetti, *Abaco Degli Edifici nel Parco del Ticino* (Alinea Editrice, Firenze, 2008), pp. 12–31
8. J. Servier, *Histoire de l'utopie* (Editions Gallimard, Paris, 1967)
9. C. Fourier, *La Théorie de l'unité universelle a paru primitivement sous le tire de Traité de l'association domestique-agricole, ou attraction industrielle* (Anthropos, Paris, 1841)
10. UNESCO Advisory Body Evaluation, Advisory body evaluation (ICOMOS). Fujian Tulou (China) (2008). Retrieved from http://whc.unesco.org/en/list/1113/documents/. Accessed 11 January 2019
11. A. Pavin, *The Kibbutz Movement, Facts and Figures 2006* (Yad Tabenkin, Ramat Efal, 2006)
12. D. Frattini, In Israele è finita l'era dei kibbutz. Nessun rappresentante nel prossimo parlamento. Corriere della Sera (2013), p. 21
13. P. Oliver, *Built to Meet Needs: Cultural Issues in Vernacular Architecture* (Elsevier, Oxford, 2006)
14. G. Gideon, *Earth-Sheltered Dwellings in Tunisia: Ancient Lessons for Modern Design* (Associated University Presses, Cranbury, 1988)
15. J. Hill, W. Woodland, Subterranean settlements in Southern Tunisia: environmental and cultural controls on morphology, community dynamics and sustainability. Geography **88**(1), 23–39 (2003)
16. I. Bentley, Individualism or community? Private enterprise housing and the council estate, in *Dunroamin: The Suburban Semi and Its Enemies*, ed. By P. Oliver, I. Davis, I. Bentley (Barrie & Jenkins, London, 1982), pp. 104–121
17. C.I. Ejizu, African traditional religions and the promotion of community-living in Africa. Resource document. African Traditions Electronic Encyclopedia (2017). Retrieved from https://traditions-afripedia.fandom.com/wiki/African_Traditional_Religions_And_The_Promotion_Of_Community-Living_In_Africa
18. S. Umali, Repackaging "traditional" architecture of the African Village in Zimbabwe. Urban and architectural heritage conservation within sustainability, in *Urban and Architectural Heritage Conservation Within Sustainability*, ed. By K. Hmood (Intechopen, London, 2018)
19. C. Laburn-Peart, Precolonial towns of Southern Africa. Integrating the teaching of planning history and urban morphology. J. Plann. Educ. Res. **21**, 267–273 (2002)
20. H. Nutini, With apologies to Dr. Caso. Am. Anthropol. **63**, 1100–1101 (1961)
21. J. Lockhart, *The Nahuas After the Conquest: A Social and Cultural History of the Indians of Central Mexico, Sixteenth Through Eighteenth Centuries* (Stanford University Press, Stanford, 1992)
22. H. Nutini, P. Carrasco, J. Taggart, *Essays on Mexican Kinship* (University of Pittsburgh Press, London, 1976)
23. M. Aguilar-Moreno, *Handbook to Life in the Aztec World* (Oxford University Press, New York, 2006)
24. R. Van Zantwijk, *The Aztec Arrangement. The Social History of Pre-Spanish Mexico* (University of Oklahoma Press, Norman, 1985)
25. M.E. Smith, *The Aztecs*, 2nd edn. (Blackwell Publishing, Oxford, 2003)
26. W.L. Merrill, M. Heras Quezada, Rarámuri personhood and ethnicity: another perspective. Am. Ethnol. **24**(2), 302–306 (1997)

References 91

27. W.L. Merrill, *Raramuri Souls: Knowledge and Social Process in Northern Mexico* (Smithsonian Institution Press, Washington, 1988)
28. W. Milliken, The Yanomami are great observers of nature. Survival International (2019). Retrieved from https://www.survivalinternational.org/articles/3162-yanomami-botanical-knowledge. Accessed 15 May 2019
29. M. Kawharu, Environment as a marae locale, *in Kaitiaki: Maori and the Environment*, ed. By R. Selby, P. Moore, M. Mulholland (Huia, Wellington, 2010), pp. 221–239
30. M.R. Austin, A description of the Maori Marae, in *The Mutual Interaction Of People and Their Built Environment*, ed. By A. Raport (Mouton Publishers, The Hague, 1976), pp. 230–241
31. P. Tapsell, Marae and tribal identity in urban Aotearoa/New Zealand. Pac. Stud. **25**(1/2), 141–171 (2002)
32. T. Amoamo, T.T. Ōpōtiki, The complementary of history and art in Tutamure meeting-house, Omarumutu Marae, Opotiki. J. Polyn. Soc. **93**(1), 5–38 (1984)

Chapter 4
Co-housing

4.1 We Still Need to Live Together

In the contemporary decades, these forms of living together are typical of the contractual communities, which were described at the end of the second chapter. Among all the typologies of contractual communities, the research focuses on the contemporary phenomenon of co-housing. Co-housing is particularly interesting because it allows the rediscovery of strong social relations and the value of sharing among a community, but also because it concerns several contemporary issues, such as resilience and fragmentation, which are extremely relevant in today's territories.

The need to live together is something innate in the human being: the participation in a social life, the security given by a social network and the need to be recognized among other individuals are aspects that are definitely innate in our being. In traditional living, these needs were met by spaces that a centuries-old building tradition had made suitable for hosting the social environment that each culture needed. The fact that modernity, with its technological thinking and its unprecedented design proposals, has redesigned living spaces in the direction of individualism does not mean at all that the needs of communities have changed. Indeed, they have been preserved and, where repressed, they have turned into new social problems.

With the development of perspective techniques, humanity began to find mathematical and geometric laws to represent reality in ways that seemed much more truthful than the previous representation techniques could do. At that time, the thought was that, if the world could be redrawn by means of these mathematical–geometric laws, it meant that the world itself was based on these mathematical–geometric laws. And if the world was based on these laws, then everything had to relate to these laws: it would not have been possible to think of something ideal and fair for man if it did not respond to these laws. Thus, starting with the Italian Renaissance, the design of ideal cities based on mathematical–geometric rules and on perspective began. Famous is the painting of the ideal city by an anonymous artist, which

© Springer Nature Switzerland AG 2020
E. Giorgi, *The Co-Housing Phenomenon*, The Urban Book Series,
https://doi.org/10.1007/978-3-030-37097-8_4

represents the perspective foreshortening of an ideal city, in which the mathematical–geometrical rules are very clear, but from which men disappeared. The city is no longer represented by people or communities, as in medieval paintings, but by mathematics and geometry, so the city is no longer designed for people or communities but according to mathematics and geometry.

This attitude toward the built environment has led, over time, to the generation of urban spaces adapted to regulations, economic evaluations and technical thought, but no longer centered around the free expression of individuals and communities. The serious thing is that the continuous perpetuation of these solutions began tarnishing our ability to understand the built environment and the spaces that human relations need. At least in the Western world, which is the main victim of this technological thought, the shared life solutions have thus gradually lost importance, to the point of becoming the minority of housing solutions. In the cities, a spatial individualism began to reign, in which the meaning of the spaces of shared life were confined only to the needs of a technical regulatory apparatus that required a defined amount of square meters for green areas, established by rules and numerical charts.

If we consider Western society, the states, proposing solutions of "social housing"—even though this is "socially" regulated by precise economic-juridical principles—tried to respond to this lack of social life in contemporary cities and citizens' housing experience.

Where the institutions did not manage to provide social supplies for the social needs of cities and territories, citizens brought forward their own initiatives. In particular, if we focus on the issues of creating a social environment around the living, then these practices are usually considered within the panorama of co-housing.

So, according to McCamant and Durrett, who considered the cultural living heritage, co-housing can be seen as a contemporary effort to regenerate those shared environments and social interactions typical of the pretechnological societies [1, 2]. The Italian architectural design studio Tamassociati, which is very involved with the topic of co-housing, believes that this phenomenon can easily be considered "revolutionary," because even if it is nothing new, it represents a change of paradigm in respect to the loneliness of contemporary everyday lives [3].

The main aspects of this revolutionary solution are the issues of sharing, sociability, and participation. The projects were born to provide new housing typologies that meet the growing demands of well-being, social recognition, fairness, and natural sustainability. New forms of neighborliness enable escape from everyday reality and the creation of new systems of collaboration and networking: co-housing generates contemporary solutions combining housing (with its undisputed privacy) with shared assets that help in socializing and promoting individual and collective activities. The possibility of having larger common areas also allows resizing, in part, of the apartments to meet economic advantages, and the grouping of economic forces, combined with the presence of wide spaces, enables the provision of additional services such as car sharing, babysitting, workshops, etc.

Notwithstanding all these benefits in terms of assets and resources, co-housing is proposing something even more relevant for the residents and for the context: this living solution is supported by a strong cultural background that brings to the cities great transformative potential.

As Guidotti underlines, when bottom-up experiences occur, intentional communities, characterized by collaboration and social protection, are created [4]. This is the aspect that interests us and that we want to emphasize: the sharing of time and space becomes sharing of benefits, skills, solidarity, and social protection.

The same book, also referenced by Mariotto, highlights the importance of the social environment created around the "traditional house" by co-housing—an environment that supports social relations and contrasts the individualistic inclinations of our contemporary society. Co-housing affects positively the mutual support and the well-being of the elderly, children, or other weak categories [5].

Moreover, as the groups are often made up of people with different skills, hobbies, or interests, co-housing can count on a richness of differences, which is an asset for the group and an opportunity for personal growth. The interesting thing, as Richard Sennett[1] said in the book *Together: The Rituals, Pleasures and Politics of Cooperation*, is that these activities of cooperation have to do with physical and social skills and they do not mean having a good heart [2].

Moreover, the relationships that govern the life of the community and the relationship between co-housers are governed not by the traditional relations between neighbors, but by the fact of being an "elective neighborhood," and by the fact that co-housers choose themselves before starting the housing adventure, and they tighten the agreements based on mutual knowing.

In summary, living surrounded by a system of social relations allows us to develop a satisfying sense of belonging to a community, but, even more importantly, it allows us to perceive a support network that can help in the case of need, and, more importantly still, it allows us to feel committed in facing problems: from the small daily troubles to the great challenges of contemporary society. This is why it is so important to talk about co-housing today.

4.2 Recent Solutions

One aspect that the research underlined in the second chapter is that, during the so-called "Industrial Revolution," state and market replaced family and community, which provided individuals with the support that they needed to live and to realize themselves. This problem of the loss of "natural" friendly bodies is certainly exacerbated when the state, through public institutions, cannot satisfy, or can only partially satisfy, individual needs. In this situation, people who can count on weak familiar networks and who are no longer supported by welfare services are pushed to recreate some communitarian structures around them. In this direction went the contemporary co-housing, which developed in northern European societies from

---

[1] American sociologist, born in 1943.

the middle of the 1970s. Original communitarian experiences were born in the 1960s of the last century in the USA and the UK with the realities of the "communes," mainly driven by ideological thinking. Those experiences, which we will not take into consideration for the research, precisely because of their strongly ideological reasons and there being some "limited cases" of sharing life, prepared the conceptual field for the development of the first cases of co-housing, and in some cases they have also been the reference experiences for the first co-housing realities. The original meaning of "co-housing" comes from the Danish translation of "*bofaelleskaber*," which means "living community" and represents the original spirit on which the first co-housing experiences were based.

The historical development and spread of this contemporary form of living together can be divided into some phases. Here will be presented the points of view expressed by Lietaert[2] and Gilo Holzmann.[3] They are quite similar, but they show one main difference regarding the co-housing for elderly people.

Following the reasoning presented by Lietaert [6], we can distinguish four stages.

- The first stage represents the rediscovery of a social and shared life that materializes in the countries of northern Europe, particularly in the Dutch reality. The first episodes are represented by activists of the feminist movements that created the first women's communities (for women only) who, albeit for their radicalism and exclusivity, can be considered forerunners of the co-housing phenomenon, of the shared urban life and in particular of the "Centrala Wonen." Skraplanet is considered the first case of a co-housing project (1972). It is the work of the architect Jan Godmand Hoyer whose idea, it is thought, was inspired by reading a psychology paper entitled "Children should have one hundred parents" [7] by Bodil Graae, expressing favorable considerations for a new type of social structure that questioned the traditional family and valorized the idea of community and sharing. Hence various experiences developed in the countries of northern Europe, with different features depending on the country and on the context. In some of them, such as Sweden, the projects are generally characterized by a strong intervention of the public sector and a vertical development, in condominiums or complexes of condominiums (emblematic is the case of Stoplyckan). In countries like the Netherlands and Denmark, projects are based more on a complex of buildings of one or two floors for a few families who share common spaces usually grouped in a dedicated building, called a "common house" (Wandlemeent and Munksoegard are characteristic examples).
- The second step in the co-housing evolution is a phase of diffusion in English-speaking countries, especially in the USA, thanks to the studies proposed by the architects McCamant and Durrett[4] [1, 8] in response to the need to create a sense

---

[2] Matthieu Lietaert, a Belgian doctor of political sciences, investigated the topic of co-housing.

[3] Gilo Holzmann, an Australian architect involved in the communitarian reality "Nourishing Mountains."

[4] They founded the "The Cohousing Company" and they are the authors of some projects that we will see, such as Doyle Street, Frog Song and Temescal Creek.

of community in a very individualistic reality. Actually, there are realities combining lifestyles related to the natural environment, such as Pine Street or realities that develop in urban contexts, often recovering existing buildings and reconnecting the urban fabric, such as Swan's Market. It should be noted that cohousing is embraced previously by those English-speaking countries that, in previous years, had already had the experience of shared life, in particular with the realities of eco-villages. In Australia we have examples like Mora Mora, and the likes of Earthsong in New Zealand, which are born with clear environmental objectives.

- The third phase sees a return of co-housing to Europe, where the success of the shared experiences of housing for the elderly can be registered. The problem of old age is a serious problem in European society and politics are always looking for solutions that could reduce the public expenditure on assistance. Although it faces a particular category like the elderly, who need assistance and care, this kind of co-housing should not be seen as closely related to the experiences of nursing homes or retirement homes. As reflected by the Franco-German film of 2011 *Et Si On Vivait Tous Ensemble?*, often older people refuse to accept themselves the limits and constraints in which sons, society and medicine want to enclose them for the last years of their lives. The film tells of five elderly people, friends for over 40 years, who decide to live, despite their problems, emphasizing the aspects of life that still excite them (the grandchildren, the dog, philosophies, photography, and women), helping each other to overcome each other's problems. The film is a hymn to the dignity of older people who, even if medicine and society take away their identity by considering them only as problems, want to still feel alive, enhancing their features and the will to live that make them more similar to children than to adults. So the group of friends found in their shared life ways to enhance their status, defending themselves from what society imposes. The "small community" of five friends is a new reality, so in the movie it is called a "trend" by a professor of ethnology. In northern Europe, in fact, many co-housing facilities for seniors have developed, encouraged by public intervention to provide answers to the issues related to the disintegration of the family network, family mobility, the lengthening of life expectancy, etc. Thus in the Netherlands, which has more than 200 co-housing communities for the elderly, we have examples like Nieue Wede, which since 1990 has been home for over 40 co-housers who share spaces, initiatives, and life moments. In Barnet, England, the community Owch project, the brainchild of a group of older women, is being developed. Lietaert argues that this development of co-housing for the "elderly" is based on the principle that it is "better to have a close acquaintance than a distant relative" [9].
- The fourth spreading phase regards the development of this phenomenon in other European countries, especially the Mediterranean ones. Since the beginning of the new century, the increasing attention given by the market, by public opinion and by the media has endorsed this phenomenon. In fact, this phase can be related to two main different realities. (1) The first reality refers more to the experiences of northern Europe, where families manage the projects independently—from

the formation of the group of co-housers, brought together by the shared interest in experiencing a renewed way of living, to the development of the design, to the realization of the project. Usually, this kind of reality is supported by associations interested in a sustainable way of life and in environmentally friendly approaches. These realities are usually related to cost saving because the needed support services are fewer (design, founding, construction, development, management, etc.), but there are considerable extensions in terms of the time required to implement the project. (2) The second is linked to groups of real estate that, similarly to what happens in the USA, offer to the families the project already realized or support the families with already experienced structures and skills to realize the project: real-estate companies, architects, facilitators and other experts support the groups of co-housers in developing the projects. These companies are attracted by the opportunity to offer on the market a new residential solution, which refreshes the traditional supply.

As previously mentioned, very similar to this analysis by Lietaert is the analysis that Gilo Holzmann carried out in 2009 about the development of the phenomenon of co-housing. He also proposes the same time phasing, excluding, however, the third phase of the development of co-housing for the elderly. His is a classification based on the geographic spread: between the 1960s and 1970s there were the "pioneering" realities of Denmark and of other Scandinavian countries; in the early 1980s, the phenomenon shifted to the USA, and finally from the early 1990s to the states of central and southern Europe and to English-speaking countries [10].

4.3 From Communitarian Micro-Dimension to Social Macro-Innovations

To better understand the role that co-housing plays in contemporary society, let us look at some of its behaviors and effects, from the sociological perspective of functionalism, or a functionalist perspective. Since this perspective makes it possible to see how the members of a society adapt to changes in order to maintain a balance of the complex, applying this sociological perspective to our study will allow us to better understand the role that co-housing projects and communities play in the contemporary rough society.

The functionalist perspective is one of the main theoretical perspectives of sociology and is based on the concept that all the aspects of a society, from institutions to social facts, collaborate with each other, helping to maintain the overall balance of the system. Thus, if an institution (such as the welfare or the education system) or a social fact (such as religion or law) changes its role in society, other social aspects must in turn be modified to keep the social structure in balance. At a time like this, in which, as we have abundantly described above, social changes are the order of the day, it is very important to understand which social actors can contribute to maintaining the social balance and how they can do it.

4.3 From Communitarian Micro-Dimension to Social Macro-Innovations

The basic needs of the individual have not changed over time; on the contrary, it is the environmental conditions that can meet these needs that have changed. On a macroscale, we could reiterate how the role that the family and the community had in providing protection and assistance to an individual was partly supplanted by the role of the state and the market. I say "partly" because state and market can provide protection and assistance, but with a different quality from what family and community could do, in particular as regards the recognition of the singularity of the individual. This is just one aspect to exemplify how co-housing finds its key role in balancing the social changes of the contemporary era. As we will see, these changes not only refer to the relationships between individuals, but have a much wider breadth: from environmental sustainability to the impact on the market (those aspects that in the functional perspective can be called "latent functions").

With this perspective in mind, the following observations are some reflections on the role that the co-housing communities have in contemporary society. The idea is that these reflections can facilitate the understanding of the innovative scope of co-housing, which is not restricted within the limits of the community well-being but propagates also to the local context. The description of the co-housing projects presented will be focused on the peculiarities and originality of the single communities. However, by reading them according to the functionalist perspective, the role that each group plays in maintaining the balance of the wider society is clear.

4.3.1 Contemporary Territories

Co-housing represents an interesting phenomenon that provides answers to a reborn unsatisfied need for community and social security. However, the interest in this new phenomenon is not limited to these aspects, as its innovative nature also concerns other fields, including its relationship with the city, its "resilience" and its existence as a cultural expression.

Co-housing is often seen by its detractors as a phenomenon that develops from ideological positions, which denies the value of property and individuality and that brings with it the risk of breaking the urban fabric, creating residential islands, closed in themselves, that do not communicate with the urban context. If, as regards the ideological and private property that we have already discussed previously, coming to the conclusion that in the successful cases the ideological component is absent and that property and privacy are guaranteed, let us now address the issue of the isolation of communities.

The issue concerns the risk that the reconstruction sense of community and collective responsibility could deepen the fragmentation of the cities. The growing phenomenon of the "gated community," which is widely discussed in this period and is often defined as a "risk," goes in the negative direction of creating a golden enclave closed to the city. It contributes to the fragmentation of the urban fabric, to the social separation and destruction of every possible phenomenon of economic, social, and cultural interaction. We can make several criticisms (in terms of economic, social, cultural, and environmental sustainability) already on the communitarian level within the gated community, where the iterations among residents are carefully avoided to ensure maximum individuality of life. If, moreover, we think about their interactions with the environment it is clear how threatening and harmful they are for the reality in which they occur.

The co-housing communities, on the other hand, start from completely different assumptions and they approach the context with methods that may vary but that are certainly different from those of the gated communities. We saw that co-housing is based on the principle that people who take part fully agree with the idea that sharing is a fundamental instrument to regain a decent standard of living and drive out the danger of alienation that the contemporary society brings. Co-housing is based on sharing (and not just on sharing within the community). Indeed, there are numerous cases in which the idea of sharing is promoted to the neighborhood, trying to accomplish tasks that can involve the whole local community. Sometimes the same architectural co-housing projects start from the definition of specific spaces that host activities for the neighborhood or other spaces to open the community to the social context in which it is inserted. The reasons may be different, from spaces for productive activities such as home-working or co-working, to spaces for the sale of produced material, to small workshops for repairing or promoting the practice of reuse and recycling, to simple multipurpose rooms for shared activities.

Fragmentation Is a Serious Problem

The perception of contemporary cities is the result of a sense of identity that we develop as a result of an assembly of fragmented experiences, the result of a territory organized by compartments through a discontinuity of spaces and places. Contemporary cities are also the result of wrong urban decisions and improper relational strategies, where too often the urban and architectural interventions contributed to increasing the fragmentation of the city and the isolation of the citizens.

Too often self-referencing projects have been produced. So the result is not a city made by exchanges and relations, but an archipelago composed of little islands and cells that cannot be compared to the neighborhood.

We are sure that, thanks to specific residential interventions, this trend will change. We are sure that we will think back to the city where man can return to being a citizen, a city that will come back to live relationships and exchanges.

If considered in a correct and complete process, co-housing becomes a residential intervention capable of creating social situations, of generating open, live and collaborative spaces and of realizing places as creators of community and exchanges.

We believe that co-housing is an important practice because of its ability to deal with the issue of fragmentation of the contemporary city [11]. This consideration comes from the fact that co-housing, by its nature, organizes the social structure in which it is set up with a new form of rediscovered relations and cooperation between citizens. In particular, the research stresses how this innovative organization of the social structure takes shape through four different levels of sharing.

- The first level refers only to the contractual aspect of the social relations. The organization of a contractual community, such as that of co-housing, is by its nature on the basis of a contract shared by the inhabitants. It essentially defines the social relations and the structure of the community itself. Of course, considering just contracts is not enough to create a real community and the research does not focus on this issue.

- Communities are based on functional relationships within the group, bringing us to the real dimension of sharing. We have seen that they may have different characteristics and the shared spaces can vary depending on the project, but they are those strictly related to the activities of living and those related to the "additional" activities of living. Often, among these "extra" activities we can find activities that are offered not only to the narrow co-housing communities, but also to the wider community of the neighborhood: to the social context.
- The co-housing community that comes to life around these shared spaces can open itself to the neighborhood. In fact, another step that characterizes almost all the contemporary experiences of co-housing is the presence, within the complex, of functions that are not reserved just for the residents, but that are opened up, in a more or less significant way, to the neighborhood. In order to promote their goal of connecting with the district, the spaces for the development of these activities are generally designed on the ground floor of the buildings. Typical examples of hosted functions that become a focal point of the neighborhood, helping to strengthen the social structure, are exhibition spaces, co-working or microcredit, time banks, gardens or spaces for training and job placement. These features, designed both for residents and for the district, may be organized by the community, or they may be developed as a form of space given to social companies or to local associations dedicated to prevention and care. With this third level of opening up to the city, it runs the risk of closing in on itself and increasing urban fragmentation. On the other hand, co-housing that is based on these principles can be considered good practice for the wellness of the whole city.
- A fourth aspect in facing the issue of fragmentation through co-housing can be highlighted. The design can go beyond the functional and social aspect and concern the innovation in the architectural field. When we intend to act on the level of the architecture, for co-housing projects, we intend to consider some design aspects, first of all to dialog on different scales with the context surrounding the project area and establish relationships with landscape and environment.[5]

4.3.2 Reuse

The revival of Western cities is based today on the considerations regarding the issues of renovation and reuse that, albeit with different meanings, indicate a more and more common design approach. The architectural culture should be directed

---

[5] If we take as reference northern European projects, the design of the ground floor takes on a relevant meaning in hosting public functions that remove barriers and reconnect urban structure.

toward the reorganization of the existing heritage and must reflect on the value of the stratification. Combining the good practice of sharing housing with the equally good practice of reuse and regeneration can only present itself as a winning combination for the challenges of the coming times.

The symbiosis between the two issues is not only the witness and the sign of the historical moment we are going through, but it is a conscious cultural and social choice [12]. The signals cast by the actors of the building process are clear and are partially received by practices and regulations, sometimes too unstable to be able to make a solid programming, which triggers an awareness of issues related to regeneration, reuse, and collective living—a union that combines the three paths to sustainability. It acts on the environmental level by requalifying and improving the use of space, on the social level by responding to the basic and the relational needs, and on the economic level by increasing the actions of renovation and reactivation of micro- and small businesses. From this first and guided reading, a series of consequences that act on issues such as security and quality of spaces can be triggered.

The exchanges between the housing practices and practices of reuse of the existing assets will increasingly come together and create actions for architectural interventions in the coming years. The two topics can be well compared for a number of reasons: for the research of pioneer interventions that informs about the new practices and the new modalities of living spaces; for the need to have more physical contact with the city, solvable with the redevelopment of abandoned spaces; for the failure of functional options that until now have been proposed in similar operations; for an in-depth research of the destinations and of the functions to be set up; for an opening up to the surrounding public and private spaces; for architectural and spatial reasons; and for a better definition of usable space.

4.3.3 Resilience

One of the themes that is now the subject of more attention from the scientific community is the concept of "resilience." It is thought of as a response to crisis situations, but given the luck and the rapid spread that this term has had, it is better to provide a brief introduction. The word "resilience" invoked as a universal remedy in the recent economic crises and interpreted in sociology as the art of mediation in conflicts and thus as social resilience in situations characterized by difficult challenges, is now held in high regard in the field of architecture, city planning, and landscape design too, having become one of the prime objectives in operations carried out following dramatic natural events, in particular those produced by the action of water [13].

The term "resilience" comes from the field of engineering and indicates the property that certain materials possess to regain the original form after they have been subjected to the stress of deformation; it is the exact opposite of fragility.[6] This

---

[6] In a comparison of engineering materials, materials such as cast iron support weights even greater than steel, but they are much more fragile in the sense that failure occurs suddenly without any previous deformation.

4.3 From Communitarian Micro-Dimension to Social Macro-Innovations

concept was later introduced, in the 1970s, in the ecological sciences through the researches of the ecologist Crawford Holling, who refers, with the term "resilience" to the ability of a system to absorb a disturbance and to reorganize itself during the changes that occur, keeping the same functions, structure, identity, and feedback. So even if the system can assume a different "state," its resilience means that the vitality of the structures and functions remains the same [14]. Referring to a system, it means therefore its ability to adapt to changes while maintaining its integrity, purpose, and identity.

According to these observations, and considering the term "resilience" applied to housing, we can extrapolate some key aspects that characterize the resilience of housing solutions. The following is a list of aspects, related to the concepts of *resilience* and *housing*, extracted from a literature review of the theme and from the re-elaboration of a previous research by the author [15].

There are five main fields that describe this relation:

- *City:* (a) integration in the local context and interconnections with the social, natural, and productive systems; (b) innovation in the urban processes; (c) diversification; (d) adaptation to the changes of the neighborhood; (e) involvement of the community; (f) giving identity to the context.
- *Risk management:* (a) creative uses and diversification; (b) redundancy of functional components; (c) assessment of environmental and social impacts; (d) awareness of hazards and vulnerability; (e) assessment of the hazards.
- *Environment:* (a) dynamic systems for the auto-production of energy; (b) local production of resources; (c) integration with natural ecological systems; (d) measures to preserve biodiversity; (e) management of the environment as heritage.
- *Community:* (a) learning and sharing skills; (b) incorporating changes; (c) management of collaboration; (d) conservation and enhancement of focal points; (e) contributions to territorial management.
- *Building:* (a) management of the impact on different scales; (b) management of building temporality; (c) sustainable impact of building material and technologies; (d) spatial flexibility.

The suggestion is to bear in mind these features while reading the projects presented in the next chapter. On average, the projects presented (the buildings and the community) will show these features with a concentration higher than typical housing solutions. This demonstrates the reason why the practices of co-housing are the most attractive in times of crisis, offering the ability to adapt to the changes of the surrounding environment, while confirming their identity. Moreover, it is interesting to note that these forms of sharing (not only co-housing but also co-working, co-production, and participatory design) often arise for reasons imposed by the need to adapt to crisis situations (savings money), but then they present themselves as a solution to the same condition of crisis, proposing solutions, ideas, and opportunities for the community and for the context.

4.3.4 Designer

In the contemporary housing reality, unlike the traditional solutions, it is logical that the role of the designer changes. The role changes not because the designed object changes, but because the way of designing changes. A key feature of co-housing projects, as we have said, is the participatory process leading to the formation of the community group.

The future residents, or the real estate company that seeks to create a group before proposing a project, are the organizers of the project. This essentially means that the design of living spaces and the programming of future management always start from a discussion between the co-housers. These discussions can be considered rich because of cultural, professional and political differences, but they require compromises and certainty that everyone is convinced of the choices made. It is not a coincidence that, if in the literature we can read only a very few cases of failures of co-housing projects, conversely we can register significant cases of projects that never see the light because people cannot overcome this delicate initial phase of solving problems and enhancing diversity. So it is important that those professionals called to support the groups designing technical and professional solutions are able to demonstrate particular abilities and skills of listening, mediation and enhancement of the ideas that emerge from the discussions. In these cases, the designer has to show even greater capacity to perceive the essence of the environment where to build and the needs of the inhabitants; he must be able to make feasible the ideas that emerge and carry them into the field of the real. On the one hand, he increases the spectrum of the possible of design solutions; on the other hand, he avoids the dreams too far beyond the possibilities of realization. The designer is therefore not only the one who has an idea and draws it, but he must also be able to synthesize the ideas of customers in a more complex way than any traditional project. It is not a coincidence that in the realities of participatory planning, in fact, recourse to community facilitators, who help to reach consensus on the choices, is needed. Also, helpful decision tools can take the most varied forms: from palette to express the index of appreciation of the proposals, to post-it to more or less articulated working groups. Every group of professionals who are called to carry out a project can develop their own organization strategy.

The designer becomes a mediator, a simplifier that finds a unique formula to bring together tastes, availability, trends, and different desires. We have already mentioned the idea of Giancarlo De Carlo, who described the participatory processes as contradictory and changing, because, unlike authoritarian design, the designers must be more receptive in understanding the client's point of view and in turning it into the starting point.

4.3.5 Real Estate

Nowadays, the situations related to the construction market represent a serious issue for the economy, both for the people who come to the market to satisfy their need for a home and for the construction companies that have been facing major problems in recent years, due to the common problem of the "unsold." Co-housing undergoes both realities, proposing itself as a way to overcome some of these difficulties.

- Except for some special cases, the choices in a traditional real-estate market make it possible to choose between the proposals already made, so that the decision is almost always motivated by economic convenience, rather than desires and dreams. The solution of co-housing is able to expand the housing supply because, thanks to the existence of a group of buyers, people can have higher purchasing power and try to get one that they wish to lease. So it is easier for a group of buyers to purchase a whole building to be restored in the city center than a rural complex immersed in a natural environment. This is certainly not a thing to be underestimated when the investment in a house is probably the biggest investment that one has to pay during one's lifetime and this must really be able to satisfy one's desires in order to feel fulfiled. This is a right. It is not just the location and the context, of course, but also the forms and contents of the complex that are relevant. With co-housing, buying the physical good of the apartment also has the benefit of "buying" a good such as the neighborhood: you choose your neighbors and you have the chance to live in an active neighborhood network based on sharing and on social support. We said that co-housing facilitates the realization of this principle because it is based on the purchasing power of a group, not because it is cheaper than the traditional supply. In fact, cost saving is not the main reason behind the decision to start an experience of co-housing; if it is true that many fixed costs come together and are shared and the size of housing may be reduced in the presence of common areas, these common spaces represent an important core cost in the economic planning.
- For the construction companies, co-housing projects are an advantage not only because of the demand but also because of the supply. In fact, the construction companies that are facing the weighty problem of the unsold may find an important ally in the project planning. As we have said, the essential starting point of a co-housing project is the presence of all the residents in the planning stage. Having the future buyers already defined ensures that the problem of the unsold will not exist or, at least, will be severely limited.
- However, the real-estate market does not seem to be so keen on the theme of co-housing: demand is focused on more traditional solutions, because there is a "cultural" tendency of the demand to turn to individual dwellings and because the supply is missing in offering nontraditional solutions. There is still a cultural component, which relates co-living to ideological obligation and memberships. So co-housing often embodies the antagonistic dimension on the market [10]. However, co-housers must be seen as "economic entities," with purchasing

106 4 Co-housing

power awareness, who aspire to impose their choices without an ideological approach, but with a strong sense of practicality and awareness, thereby defining a new type of demand that is increasingly complex and organized.

4.3.6 The Reverse: Gated Communities

It should be clarified, of course, that not all contractual communities aim to create a community, or to reduce the isolation of the contemporary individuals. Contemporary cities are experiencing a period of intense transformation that comes along with essential challenges for urban territory regarding social equity and environmental well-being. One of the phenomena that contribute significantly to the social and environmental aspects of crises in the cities is the so-called Private Gated Communities (PGCs). PGCs are areas of cities closed to the urban public life, where the owners hire companies to provide the main services. During the last few years, this phenomenon grew up mainly as a security solution, but this new wave of construction creates issues with accessibility, traffic, and social and fiscal boundaries—unwalkable cities. Having arisen and developed in the USA (it is estimated that today in the USA 5% of the population are living in these forms of reassuring isolation), now they are spreading into the European world. In the northern part of the Mexican city of León (city in the state of Guanajuato with 1.7 million people), 30% of the urban area is closed by PGCs [16], whose dimensions vary from several hectares to several square kilometers. The sprawl of the Mexican cities, combined with the disconnection generated by these PGCs, produces urban territories where the commute is entirely dependent on private cars since urban points of interest are at great distances from each other. These distances are often difficult to cover on foot, which resulted in roads that are mainly designed for cars. In terms of the dimension of the phenomenon and the cultural differences that influence the urban developments, PGCs show very dissimilar characteristics, even if they are united by the aim of giving a stronger feeling of security to the users. As a result of this aim, sometimes the gated communities risk turning into golden prisons, where contacts between neighbors are limited and privacy becomes the organizer of the users' lives, who find solace in the technology that watches over them.

Without spending too long on the interesting subject of PGCs, which deserves a much more in-depth discussion to understand the reasons for their existence, their characteristics, their political-contractual role and their impact on the equity and health of the territories, there is an observation to be made regarding the main difference with co-housing. This reasoning is worth mentioning because often people tend to equate co-housing and gated communities for the simple reasons that both are defined as "communities" and that both are well geographically outlined. However, to differentiate the two realities there are many elements to consider, all caused by this basic difference: the purpose for which these two realities are generated. On the one hand, PGCs want to offer on the market privileged housing solutions that, by isolating the housing from the existing context, respond to different possible needs of security, quality of services, elitism, etc. On the other hand, co-housing communities want to offer to their members, who are generally the creators of the project itself (thereby excluding the passage of market logic), a friendly environment where it is possible to generate and strengthen social relationships of mutual support, without the need to isolate themselves from, or close themselves to, the existing context. This main difference brings to others different approaches to living: the presence and dimensions of common facilities, the way in which shared spaces are used, the activities performed and the relations with the surrounding context.

4.4 Common Characteristics

As just shown, almost none of the co-housing realities that have been developed in recent decades have been motivated by extreme ideological reasons, as happened in the communes of the 1960s and 1970s, but they are usually promoted by the search for a better quality of life. In fact, while the urban contemporary realities attract because of the many possibilities they offer, at the same time these new realities intimidate due to the difficulties of living in conditions of forced individualism and weak social ties. This leads to a loss in the sense of fulfilment, security, and satisfaction. So it is for this reason that people look carefully at what co-housing offers. More and more often, we see the same people who put themselves in the front row to face the problem: they form associations, define the community and begin the process for the construction of their co-housing. In almost all cases, in fact, projects start from the same people who will live in the project, which is designed from the first stages as tailored to the residents. For this reason, co-housing projects enjoy great success. Co-housing, in fact, allows desires and needs to be met, combining space custom-designed in an important phase of participatory planning, in strong and healthy neighborly relations.

With the aim of clarifying which characteristics can define a co-housing project as such and which can describe a co-housing project, we first present a general consideration about the phenomenon, as can be understood from the study of ongoing experiences and of the existing literature. The main reflection that I consider essential for approaching this topic concerns the fact that the term "co-housing" refers to a significant variety of very different projects, involving countries with different social realities, needs, and cultures that lead the inhabitants to develop responses that may seem very different from one another. For example, if we feel that the starting reasons may be driven by social needs, such as social security or a renewed sense of community, rather than economic or environmental ones, we understand that the range of project outcomes is indeed ample. Moreover, the design process that contemplates high levels of flexibility means that each project is different from the others, because it is designed by and for the people that will live there. As the design is flexible, so the management of the community is elastic: co-housing communities can never be rigid communities, with external imposition of rules, since the co-housers define the regulation of their own "village." In a co-housing community nothing is rigidly fixed: the dimension and the function of the common spaces, the internal organization, the activities, even the new members, etc. are always decided—albeit with different decision-making methodologies—collegially. The possibility of conciliating the respect for one's own privacy with a life in a shared communitarian environment is a huge wealth that today a lot of people are looking for. If co-housing projects go beyond entrenchment in one's own apartment, revealed as being more complex than the traditional housing solutions, we must not overlook the fact that they preserve intact the private dimension of living, which is not sacrificed by the sense of space and sharing of moments that are just added value to the quality of life. This is exactly what can explain the success of co-housing at the beginning of the twenty-first century: the possibility of choosing how and with whom to live and dwell, sharing one's everyday life and the choices in the organization of it.

The recent reality of co-housing records attitudes mainly marked toward environmentalism that can be seen in several issues: eco-design and green-techniques construction; the eco-friendly management of communities; the use of local resources; local production and consumption. So it is not just a matter of using green building techniques for construction, or adopting sustainable design features, but seeing sustainability as a whole. Even planned activities will be sustainable in the sense that they will try to integrate the residential function with production, which can range from the partial auto-production of food, to the creation of laboratories for DIY and home repair, to activities to encourage the practice of recycling and sharing or production of alternative energy. All these initiatives, overall, are often considered in relation to the social and economic reality of the neighborhood that the communities interface with. So this is a sustainable approach to save resources, to create sustainable alternatives and networks to share within the community and with the neighborhood, which is the real recipe for environmental sustainability (social and natural). In fact, it is usual to see communities that combine the housing functions with labor activities, which allow exchanges of knowledge and skills within the community.

Finally, an aspect that must never be forgotten and indeed stands as one of the key features of the projects is that they should not be imposed. Co-housing should not be imposed and this was told to us in the first decades of the twentieth century by Moisej Yakovlevich Ginzburg[7] talking about the Strojkom. Referring to this project, the Russian architect underlines how it is impossible to make the collective use of a building mandatory. On the contrary, designers should create spaces that facilitate a gradual transition to the use of common facilities and allow a natural blooming of shared activities [1]. However, for people interested in the subject of design and planning, the aspect that seems most relevant to me, among all those that emerge from the discussions on co-housing, is that concerning the role of design, because a good architectural project is basic to generate profound internal and external

---

[7] Moisej Yakovlevich Ginzburg (1892–1946) was a Russian architect.

relations; in the same way, the process that leads to the creation of strong and lasting social bonds is fundamental. This process starts from the first phase of setting up a group for the management of the community itself. Therefore, design for co-housing does not stop at spatial planning, but goes further, in an overall vision of the process of creating and maintaining the community.

4.5 Six Features to Define "Co-housing"

According to what has been said, here is a list of features that can be considered requirements to define a communitarian project, a project of co-housing.

1. Participation in the communitarian life

   The residents actively participate in the life of the community. From the starting phases of the development of the project to the decisions made, dinners, working, relaxation, and other common activities become good opportunities to create that sense of sharing that is at the base of the idea of "co-housing."

2. Wide common issues

   The everyday sharing of activities, experiences, and time is at the base of the idea of co-housing; to do this, wide common spaces, and integration of private spaces, are required. Depending on the degree of "co-housing," the dimensions, quality, and functions of these spaces can change considerably.

3. Free will to participate in an original housing experience

   The participation in any kind of sharing living method must be the result of a free decision taken by the co-housers in complete freedom. Projects coming from higher plans imposed on people, even if they give a lot of attention to sharing and common spaces, cannot be considered "co-housing." The will must refer to participation in a new housing experience, where the topic of the house is central. This is a remarkable aspect because several other communities are organized around the will to participate in a spiritual or environmental experience, and for this reason, even if they must be considered communities, they cannot be considered "co-housing communities."

4. Direct management of communitarian issues

   The residents take care directly of the spaces and the management of the community, in particular in regard to decision-making.

5. Nonhierarchical structure

   The strong time–space sharing in the co-living of different families and people requires several moments of decision-making. These decisions must be taken by the whole community, without resorting to any form of hierarchical structure.

6. Non-paid common activities

   The activities and works that are required of the co-housers to keep the buildings efficient cannot become an income source for any of the residents.

4.6 Three Main Differences

These are the characteristics required for a project to be considered a "co-housing" project. This effort to find some common characteristics that could define a co-housing project is due to the fact that the panorama of the contractual communities considered co-housing is very wide. The way in which people live in a shared environment depends on a lot of factors, which can mainly be related to these three categories: (1) the reasons why people look for this kind of shared living; (2) the level of sharing within the community; and (3) the context in which the community is inserted.

1. The reasons

   The reasons why a group of people decide to live together vary according to the local culture, the sensitivity of the community toward particular items, personal interests, the availability of places or the desire for a particular kind of living environment. Within all these differences, which actually are very specific for each case and vary a lot from case to case, there are at least four general tendencies that can be outlined: (a) the desire for security: to live in an environment made secure and protected by a community. This case refers to projects in which the co-housers try to create a social environment capable of protecting the inhabitants, particularly those weaker categories such as the elderly, children, and the disabled. Sometimes the community can be composed of people belonging to the same category, as in the case of co-housing for the elderly, or of a mixture of people, promoting mutual support; (b) the desire to connect with a natural environment, far removed from urban life, in which to rebuild relationships with nature and people. Usually this tendency passes through a more contemplative relationship with nature, to be performed within social groups or by individuals, or through agricultural production activities, which foster to an even greater extent the social bounds within the community; (c) the desire to rediscover the social dimension of living, escaping the individualistic way of contemporary living to create a new social environment, capable of offering opportunities for connections, support and sharing of material and ephemeral resources. This desire is connected to the most intimate reason for co-housing, and often, in this case, the community works as a promoter or spreader of these new social dimensions, thanks to the involvement of the local external communities. (d) the desire to live in an environment in which there is an opportunity to create common activities that develop around shared services, complementing life (library, sauna, gym, etc.). Co-housing implies sharing of activities and spaces, but in this case, the sharing develops around something not central to living.

2. The levels of sharing

   The levels of sharing indicate the threshold at which the inhabitants are willing to share activities, spaces, and structures with the other members of the community. It is known that every person who is part of a co-housing settlement has their own personal space, in which the necessary privacy is guaranteed. This means that there is always a private dwelling for each family, usually with everything needed to live independently (kitchen, bathroom, and bedroom). What varies are the shared spaces and resources, as well as the planned activities for community life. As will be seen in the cases, some of them share only services complementary to living (bookstore, gym, discussion center, gardens, etc.); others share more intimate services (kitchen, living room, workshops, etc.) that allow you to carry out activities more closely related to the essence of living (cooking together, eating together, talking openly to each other, deeper support, etc.); and others share services for productive activities (workshops, orchards, small livestock, etc.) to enable activities that strengthen personal relationships and allow the community to count on local production, which can reduce the environmental impact, or be a source of economic savings or even of revenue. In some co-housing, the level of sharing is even more stringent, to the point of being at a point of equivocation between being a co-housing, a commune or an eco-village. As already stated, in fact, the great variety with which this phenomenon of shared life takes place is so vast that sometimes categorizations, although necessary to better understand these experiences, become narrow and cannot be read too strictly.

3. The context

The context where a co-housing project takes place can easily vary without questioning the deep essence of co-housing: the same shared living can happen in an urban, rural or forest context. The fact of being settled in one context or another may change the relationships that the community will establish with the surroundings and even the internal dynamics of the community, but this does not mean that some projects are or are not co-housing. For example, a project that fits into an urban context will have a greater tendency to work with the neighborhood community than a project that fits into a forest context, which may be characterized by being more closed in itself. Or a rural project, in which social relations develop through production work, should not be considered more or less worthy of a suburban project in which older people share activities to stay active and strengthen a friendly social environment.

4.7 Four Categories Help in Describing the Realities

In addition to the requirements to define "co-housing" and the principal three categories that differentiate the projects, here additional secondary features are proposed, in order to better help in reading and describing the projects. So, even if they cannot be used to say whether a project is a "co-housing" or not, they want to be like a tool for reference, which helps in seeing how projects can be portrayed and in highlighting the main features that characterize the vast world of co-housing. The proposal, which comes from the analysis of ongoing communities and from studying the existing literature, groups these features into four main categories, each of which is composed of a variable number of topics:

Community, which reflects the communitarian spirit present within a community: the desire to share, the planned activities, the presence of spaces designed to share, etc.

Housing, which reflects the residential way of living, in particular with reference to the involvement of the residents in the activity of living and housing: from the planning of the project to the management of production.

Design, which reflects the design methods used to develop the project, the approaches toward sustainable practices and the participation of the future residents.

Context, which reflects the relations with the natural and social environment, according to the involvement of the community in the life of the local neighborhood.

1. Community

   a. Sharing skills: The interaction between groups of residents with different skills becomes one of the most important opportunities to develop the sense of community and increases the capability to improve the group itself and solve the problems;
   b. Feeling security: People can perceive a higher level of safety and security inside the residential community, thanks to a friendly environment; we can also refer to the fact that the project wants to create an environment particularly well protected for young and older people.
   c. Collaboration: the groups of residents develop, thanks to the interaction in the living and production activities, also scheduled, a great sense of collaboration inside the community.
   d. Decision consensus: Taking decisions concerning the lives of the community residents can develop in different ways and the consensus in terms of choice is the one that produces the highest sense of community.
   e. Shared meals: People can schedule having common meals that improve the bonds inside the group; in this way, meals are not considered just a private moment, but a good opportunity to define community.
   f. Focal spaces for community: This means the presence of common spaces that become focal places where the members of co-housing can meet or around which they can create a sense of community.
   g. Common residential spaces: The project provides for the development of common spaces that host activities that can be considered basic for living; they are the spaces in which cooking, washing or having intimate conversation; they can be in addition to the equivalent private ones.
   h. Common complementary spaces: The co-housing project has some shared spaces for activities that cannot be considered basic for living, but rather "complementary" to the basic ones.

2. Housing

   a. Mix of residents: Persons with different features, mainly regarding age and economic class, but also sex, religion, culture or ideas, make up the group of people involved. This mixture increases the development of the interaction between the members, the mutual help and the co-presence of different skills and experiences; it also distances the danger of isolation of the community from the social context.

   b. Self-development: The group of residents are the promoters of the project; to facilitate the realization, and depending on national rules, they can be organized into associations or cooperatives; clearly, depending on the rules of the community, new people can join the starting group.

   c. Self-construction: The residents participate actively in the development and in the realization of the project, thereby increasing the collaboration and the sense of group.

   d. Link with production activities: People have the opportunity to develop common activities for the production of services or goods that can be addressed to the community itself or be directed to external selling; typically, based on the environmental attitude of the co-housing organization, they have a sustainable approach.

   e. Partial food production: The members of the group produce in the common gardens part of the food for the demand of the community itself; this sustainable feature increases the bond between the co-housers and the common propriety.

   f. Privacy: The project provides common spaces for shared activities, but it guarantees private spaces for all the residents so that everybody can benefit from the necessary intimacy and privacy. If essential services such as a kitchen or bathroom are not in the private dwelling, we do not consider it private. The presence of relatively high levels of privacy is the most relevant feature that allows a co-housing project to be differentiated from an eco-village one.

3. Design

   a. Sustainable approach: The lifestyle of the community is based on a sustainable approach and all the activities and choices follow ideas of simplicity, adaptability, and solidarity.

   b. Reuse: The co-housing project is based on the reuse of existing buildings; this architectural approach allows the two virtuous approaches of sharing and reuse to be combined; it can refer not just to ex-residential buildings.

   c. Flexibility: In accordance with the importance of adapting to the different circumstances and evolutions of the community and its common need, the design is based on the search for flexibility and elasticity of use.

   d. Participatory design: Following the principles of participatory design, the residents take part in the development of the design activities, thanks to professionals who develop the project; this allows the realization of spaces very close to the wishes of the habitants and the creation of a stronger bond among the co-housers, while also testing their decision-making capacity.

4. Context

   a. Integration: The project is integrated in the local context, developing and managing the interconnections among the social, natural, and organizational systems (infrastructures, information, etc.).

b. Innovation: The presence of co-housing and its activities represents a strong innovation impact for the environment in the social and urban processes.
c. Participation: Involving the neighborhood, and developing the participation, on all scales, of the members of the community.
d. Identity: Planning the intervention to make it a focal component of the neighborhood system and to increase the local sense of identity.

These features were used as a guide to interpret the design cases described in the following chapter. Without applying them as a schematic reference, these points have been followed to interpret the projects, to extract the fundamental aspects and to describe them. For each case, a checklist of characteristics has not been made because, according to the author, this would have been too objective a process, deriving from a personal understanding of the case. Moreover, a simple check of the characteristics would be very limiting for the complete description of the project, as a feature may appear partly, in a nonconstant manner or, although present, may not have the same importance as other projects. So, we repeat, the list of these characteristics is only to allow a better reflection of the cases, allowing us to see on which aspects to focus attention when studying and describing the projects.

Moreover, it must be taken into account that this list describes the phenomenon of co-housing with a general overview of the experience, trying to touch the most relevant aspects that can differentiate the projects, and it can certainly be improved or adapted if there are any specific categories to investigate. Moreover, the proposed instrument may be considered a set of "best practices," but we must be aware that the features should only help in understanding the project, and do not represent guidelines for the design.

References

1. G.D. Manzoni, E. Giorgi, T. Cattaneo, Recover and cohousing: looking for environmental and social sustainability. The intervention modalities for dismissed shopping centers, manufacturing and residential buildings, in *Cohousing. Programs and Projects to Recover Heritage Buildings*, ed. By A.F.L. Baratta, F. Finucci, S. Gabriele, A. Metta, L. Montuori, V. Palmieri (ETS, Pisa, 2014), pp. 99–103
2. R. Sennett, *Together. The Rituals, Pleasures and Politics of Cooperation* (Yale University Press, New Haven, 2012)
3. Studio Tamassociati, *Vivere insieme. Cohousing e comunità solidali* (Altreconomia, Milano, 2012)
4. F. Guidotti, *Ecovillaggi e Cohousing. Dove Sono, Chi li Anima, Come Farne Parte o Realizzarne di Nuovi* (Terra Nuova edizioni, Cesena, 2013)
5. K. McCamant, C. Durrett, *Creating Cohousing. Building Sustainable Communities* (New Society Publishers, Gabriola Island, 2011)
6. M. Lietaert, Il cohousing: origini, storia ed evoluzione in Europa e nel mondo, in *Famiglie, Reti Famigliari e Cohousing*, ed. By A. Sapio (Franco Angeli Editore, Milano, 2010), pp. 140–148
7. B. Graae, *Børn Skal have Hundrede Foraeldre* (Politiken, Copenhagen, 1967)

References

8. A. Mariotto, Il cohousing come pratica di cittadinanza organizzata, in *Vivere Insieme. Cohousing e Comunità Solidali*, ed. By Studio Tamassociati (Altreconomia, Milano, 2012), pp. 17–32
9. A. Sapio, *Famiglie, Reti Famigliari e Cohousing* (Franco Angeli Editore, Milano, 2010)
10. S. Sfriso, Dire, fare… coabitare, in *L'abitare Condiviso: le Residenze Collettive Dalle Origini al Cohousing*, ed. By E. Narne, S. Sfriso (Marsilio, Venezia, 2013), pp. 43–59
11. E. Giorgi, I. Delsante, N. Bertolino, Collective housing as a good way to solve city fragmentation, in *Cohousing. Programs and Projects to Recover Heritage Buildings*, ed. By A.F.L. Baratta, F. Finucci, S. Gabriele, A. Metta, L. Montuori, V. Palmieri (ETS, Pisa, 2014), pp. 196–201
12. K. McCamant, C. Durrett, *Cohousing: A Contemporary Approach to Housing Ourselves* (Ten Speed Press, Berkeley, 2004)
13. P. Nicolin, The properties of resilience. Lotus **155**, 52–57 (2014)
14. T. Jackson, *Prosperità Senza Crescita* (Edizioni Ambiente, Milano, 2011)
15. E. Giorgi, G.D. Manzoni, T. Cattaneo, Resilience: co-fighting the crisis. in *Architecture and Resilience on the Human Scale. Architecture and Resilience on the Human Scale, 10–12 Sep 2015, Sheffield* (The School of Architecture, University of Sheffield, Sheffield, 2015), pp. 79–87
16. C. Charles, P. Gutierrez, E. Giorgi, H. Gutierrez, V. Barquero, A quién pertenece la ciudad? Movilización, espacio público, tejido social y bienestar: La privatización de calles en la zona norte de la ciudad de León. Entretextos **26**, 58–74 (2017)
17. E. Battisti, *Architettura, ideologia e scienza. Teoria e pratica nelle discipline di progetto* (Feltrinelli, Milano, 1975)

Chapter 5
Co-housing Cases

5.1 Studying the Cases

From the first idea of writing the book, the intention was to present co-housing through a collection of an important number of cases that could give a valid representation of the wide variety that this phenomenon can assume and, particularly for readers who approach the topic for the first time, a first-hand impression of the phenomenon. Given the great variety of forms that co-housing projects can take and the innumerable facets with which this phenomenon can occur, the choice of which cases to present has been very interesting and demanding. The intention is to give a representation of the phenomenon that is as truthful as possible, thereby presenting both the cases closest to the traditional conception of "co-housing" and those that are at the limits and that often highlight some of the new trends toward which this phenomenon of shared living moves. So, to define a list of projects that was as representative as possible, a choice was made to start with a vast number of projects, and then filter them and reach the final number of 50. The methodology for defining the final list comprised three steps.

First List
Drawing on various sources, such as academic articles and books on the subject of co-housing, Internet pages and community profiles on social networks, a list of around 300 projects was prepared. In this list there were two main types of projects: (a) about 270 cases that defined themselves as "co-housing" or that were recognized by the scientific community as "co-housing"; (b) about 20 cases that, although not explicitly declaring themselves as "co-housing" in their presentation, possess some characteristics that can make them fall into this category. They were, in particular, eco-villages whose characteristics make them eligible to be part of co-housing projects. Combining these two lists, we arrived at a total of almost 290 potential cases that have been mapped to provide an idea of the geographical distribution of the phenomenon, which, clearly, became denser in the regions of North America, Australia, and Europe, and scarcer in regions such as Africa and western Asia. In

© Springer Nature Switzerland AG 2020
E. Giorgi, *The Co-Housing Phenomenon*, The Urban Book Series,
https://doi.org/10.1007/978-3-030-37097-8_5

addition to these geographical differences, the projects also had differences in the stages of realization: some cases had already been in operation for several years, others had just been established, others were still in the process of constructing spaces and others even in the process of forming the community. Although the original idea was to collect only already established projects, in order to better evaluate the actual existing social relations, it was decided, at this stage, not to exclude projects in progress to see if they could present relevant aspects.

Second List

The 290 projects collected in the first list were studied in greater depth, to allow fuller consideration, and thus achieve a further scanning. This study allowed the projects to be skimmed based on certain aspects, including:

- Effective coherence of the project with the definition of co-housing as presented in the previous chapter. Very often, in fact, the word "co-housing" is associated with projects, very few of which deal with shared living. This tendency allows us to reflect on a relevant aspect: many developers and designers use the term "co-housing" to sell their product, taking advantage of the interest that the phenomenon has attracted within the contemporary society, which evidently already links the association of the terms "sharing" and "housing" to a sustainable solution.
- Availability of descriptive material and ease of access. Cases in which the available material was limited and showed little relevance were discarded. On the other hand, in cases when little material was available, but there was sufficient to express the relevance of the project, the community was contacted and asked to provide more information.
- Geographical distribution. In geographical areas dense with projects, those of limited interest to the dissertation were excluded to skim the highest possible variety of cases. This happened particularly in the USA and in northern Europe, where a lot of cases present similar characteristics in a limited territory.
- State of the community. The cases not yet realized, and with insufficient features to introduce new themes and interesting considerations in the dissertation, have been excluded.

This skimming allowed a second list containing only 84 projects to be created. By mapping again the geographical distribution of the projects, a situation similar to the previous one was obtained, with the exclusion of some areas such as South Africa, India, and Japan.

Final List

Using the contact information taken from web pages or on the social networks of the communities, we tried to contact all the 84 selected companies. In those cases where it was not possible to have direct contact with the community, we tried to contact the design studios that carried out the project (obviously when they were known). The fact of not being able to contact all the 84 selected communities to obtain more information or graphic material produced a further skimming that led to a list of 56 projects. Some of these, however, did not want to be presented in the book or did not follow up on the correspondence. At the end, of the 56 projects, 46 cases were taken into consideration, with which it was possible to have constant and productive communication. Another four cases, whose relevance in the contemporary co-housing panorama is very high, have been added to this list, even though there was no direct correspondence with the communities.

It seemed right to give an explanation, albeit brief, of the adopted methodology because otherwise the choice of cases could have seemed superficial and subjective. On the other hand, the fact of having been able to count on such a vast starting range of 290 cases allowed us to have a broader idea of the phenomenon and to be able to make the most impartial and representative possible skim. Despite the efforts made, there remains the awareness that: (1) in situations where similar characteristics were repeated in geographically close projects, it was necessary to discard objectively interesting cases that otherwise deserved to appear; and (2) the impossibility of accessing deeper information than that available and the impossibility of communicating with some communities led to many potentially interesting cases being excluded. My personal hope is to be able to continue, in subsequent works, to gather information about these realities that are becoming increasingly present in today's territories.

5.2 The 50 Cases

The final result, therefore, is a collection of 50 projects, chosen to give a broader representation of the global phenomenon of co-housing. There are projects whose features fully depict the definition of co-housing and others that, on the other hand, are on the margins of this vast reality, in particular because they are strongly promoted by an entrepreneurial push or because, in contrast, they are very close to cases of eco-villages. Among the selected cases some are not yet active, but they introduce some interesting topics to discuss. The cases presented are therefore a selection based on the geographical characteristics, motivation, size, and context in which they are inserted. With the process described above, we tried to give the greatest possible representation of the phenomenon. The final list is shown in Table 5.1.

All the projects are presented through a format containing some data, a text, and some support images, which in most cases have been suggested by the members of the community to best describe their personal perception of their shared-life experience. To write the projects' descriptions, the material available in the web pages of the communities has been used or an interview with a representative of the community was organized. In several cases, one member of the community or the person responsible for the design studio/construction company contributed to enriching and checking a first version proposed by me, in order to give a better explanation of the project. In this case, the name of the co-housing member is indicated in the footnote.

120   5 Co-housing Cases

Table 5.1 The final list of co-housing projects

1	Belterra	CAN	26	Mount Camphill	UK
2	Bloomington C.	USA	27	Munksøgård	DNK
3	Cannock mill	UK	28	Mura San Carlo	ITA
4	Casa tucuna	BRA	29	Nubanusit	USA
5	Chiaravalle	ITA	30	Old Women C.	UK
6	Coflats stroud	GB	31	Pacific gardens	CAN
7	Cohousing Israel	ISR	32	Pioneer valley	USA
8	Copper lane	GB	33	Pomali	AST
9	CosyCoh	ITA	34	Quattro passi	ITA
10	Cranberry commons	CAN	35	Quayside village	CAN
11	Doyle Street	USA	36	Radiance cohousing	CAN
12	Drivhuset	DNK	37	Rancho la Salud	MEX
13	Earthsong	NZL	38	Rocky hill	USA
14	Ecosol	ITA	39	Solidaria	ITA
15	Emerson commons	USA	40	Springhill	UK
16	Forgebank	UK	41	Stolplyckan	SWE
17	Frog song	USA	42	Sunflower	FRA
18	Il Mucchio	ITA	43	Swan's market	USA
19	Itaca	ITA	44	Temescal creek	USA
20	K1 project	UK	45	TerraCielo	ITA
21	Le Torri	ITA	46	Urban Village Bovisa	ITA
22	Lilac grove	UK	47	Vaubandistrict	DEU
23	Los Portales	ESP	48	Wandelmeent	NDL
24	Milagro	USA	49	WindSong	CAN
25	Moora Moora	AUS	50	Wolf willow	CAN

5.2.1 Belterra[1] [1]

Address: 726 Belterra Road, Bowen Island, BC V0N 1G2, Canada
    Setting: forest
    Status: established (2015)
    Size: 36 apartments
    Developer: Cohousing Development Consulting and Burtnick Enterprises
    Architect: Mobius Architecture
    Decision-making: consensus
    Shared facilities and features: common house with kitchen, living room, guest house and playground, garden and workshop
    Target: mixed families
    Website: www.belterracohousing.ca
    This co-housing project, although completely immersed in a natural context (Fig. 5.1), is located just 40 min away from Vancouver's downtown. In order to

---

[1] Matthew van der Giessen.

5.2 The 50 Cases

Fig. 5.1 Winter view of the five residential buildings of Belterra (credit of Matthew van der Giessen (for Belterra Cohousing))

respect this natural environment and in accordance with the philosophical approach of the community, which wanted to find a sustainable way of living, the whole project, from the design of the construction to the everyday management, is based on an eco-friendly approach. According to this vision, the felled trees became part of the wooden structure of the new buildings and the green fields around the houses host fruit trees (while chickens and bees are expected in the near future). The buildings of the community are located on one side of a hill and connected with footpaths, while a driveway goes around the complex (Figs. 5.2 and 5.3). Along this driveway, coming from the village, the firs visible building is the common house, the core of the community, which hosts the most important covered common spaces, such as the indoor kitchen with a living room (Fig. 5.4), a playroom and two guest rooms. Nearby, around this main building, other shared spaces for open-air common activities are placed (an open-air kitchen, a garden, a plaza, a covered porch, etc.). Other common spaces are located around the complex, mainly for workshops, production and agriculture. The 30 private dwellings for the families are organized in the other five residential buildings that make up the whole project. Each building offers several dwellings, different in size and distribution, but the privacy of every apartment has been accurately preserved. The constructions, now completed, stopped at the drywall primed stage, leaving the co-housers to enrich them with their final personal touches. A final note on the sustainable approach concerns the arrangement of eventual retrofitting of solar panels in all the buildings.

Fig. 5.2 Winter view of the residential building and the landscape of Belterra (credit of Matthew van der Giessen (for Belterra Cohousing))

Fig. 5.3 View over the complex of Belterra and glimpse over the forests (credit of Matthew van der Giessen (for Belterra Cohousing))

5.2 The 50 Cases

Fig. 5.4 Dining and living room of the common house (credit of Matthew van der Giessen (for Belterra Cohousing))

5.2.2 Bloomington[2] [2]

Address: 2011 South Maxwell Street, Bloomington, Indiana 47401
Setting: urban infill
Status: established (2018)
Size: 27 single-family homes
Developer: Loren Wood Builders
Architect: Marc Cornett Architects and Urbanists
Decision-making: consensus
Shared facilities and features: common house with kitchen, living room, guest house and playground, vegetable garden and workshop
Target: mixed families
Website: www.bloomingtoncohousing.org

Even though the project has not yet been realized and the community is not yet completely formed, it is important to include Bloomington CoHousing among the 50 cases because it proposes an innovative way of offering co-housing projects on the market. In fact, (1) the community, which is usually the initial element from which the project arises, will only be formed later, and (2) the participatory design that usually allows the community to create the spaces in which to live will be carried out in a first general phase by the architectural studio and by the developers, and then will be refined in detail by the residents and the community, once they are fixed. Because of these characteristics, the Bloomington co-housing is not exactly an example of "traditional co-housing" and is considered controversial by the most

---

[2] Ernesto Castañeda.

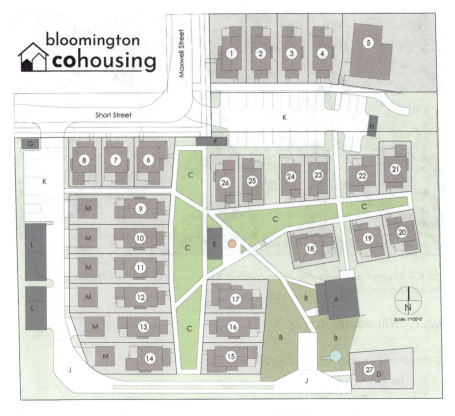

Fig. 5.5 Masterplan of Bloomington co-housing (credit of Loren Wood Builders)

intransigent observers; however, it remains an interesting example of how the market is adapting to this new form of living.

According to the Foundation for Intentional Community, Bloomington co-housing is presented as the first intentional community project in the state of Indiana that can be defined as "co-housing," unlike other cases of intentional communities of eco-villages or shared housing for students surveyed by this foundation. Also, the goals fixed by the community are very similar to those of other co-housing communities: on the web page you can read the plan "to be a community of all ages sharing the values of cooperation with one another and the environment, living peacefully together, and having fun." This plan is achieved through the organization of 27 homes combined with common amenities and a main common house (Fig. 5.5). As previously mentioned, the design of the apartments was done by the design studio and by the developers; however, by proposing customizable solutions, residents are left with the possibility of adapting the interior spaces of their homes. In the same way, even the common house is designed and built, but will leave the community, once formed, the freedom to design and define the internal spaces. A further element of aggregation for the community is the picnic shelter that will offer a

Fig. 5.6 Rendered view of the garden, taken from the external path (credit of Loren Wood Builders)

sheltered outdoor space, in a point of union between the two main green spaces of the community (Figs. 5.6, 5.7, and 5.8).

5.2.3 Cannock Mill[3] [3]

Address: Cannock Mill, Old Heath Road, Colchester CO2 8YY England
  Setting: urban
  Status: completed (2019)
  Size: 23 homes (6 flats and 17 houses)
  Developer: self-build, main contractor Jerram Falkus Construction
  Architect: Anne Thorne Associates
  Decision-making: consensus
  Shared facilities and features: open-air spaces including mill pond and common house, with kitchen, multipurpose room, dining room and utility/training room, guest rooms
  Target: singles and couples mostly 50+ age range
  Website: www.cannockmillcohousingcolchester.co.uk
  Cannock Mill is a project that respects its natural and historical heritage. It is situated in a wooded valley that, since the fourteenth century, has been the site of three water mills. This valley is now within the Borough of Colchester and the

---

[3] Phil McGeevor.

Fig. 5.7 Rendered view of shared garden and path, taken from a private house porch (credit of Loren Wood Builders)

Fig. 5.8 Rendered view of the outdoor spaces from the common house porch (credit of Loren Wood Builders)

latest mill to be constructed on this site (in 1861) has been restored to form the common house. The co-housing group is creating a supportive intentional community contributing to the social, cultural, and economic fabric of the locality. It is based on three main values: (1) good neighborliness; (2) active ageing (to encourage

5.2 The 50 Cases

Fig. 5.9 Rear of houses from shared garden (credit of Phil McGeevor)

healthy, independent and active lifestyles); and (3) eco-awareness (through building low-energy and environmentally friendly homes and sharing resources such as cars, guest rooms, bikes, equipment, etc.). The 30 residents come from different backgrounds, geographically and socially, but aim to use their varied life experiences to create an interesting social environment with high levels of trust and responsibility. There are no age restrictions, but the current members are all between 50 and 80 and the community will recruit future members with the intention of maintaining the age range and, where possible, reducing the average age. The ethos of the group can generally be described as "eco-active and dedicated to getting things done." There is an emphasis on creativity (woodwork, ceramics, and textiles etc.) and recreation (gardening, cycling, cooking, walking, theater going, etc.). The architect of the 23 homes (Fig. 5.9), built to Passivhaus standards is herself a member of the community and the group has been closely involved in the overall design and specification (Figs. 5.10, 5.11, and 5.12). The main focus of the design has been on the integration of new modern homes with the traditional mill. The mill is the common house and the base for the shared activities—cooking, eating, and socializing together (Fig. 5.13). The site also has a shared garden and growing area and an attractive mill pond.

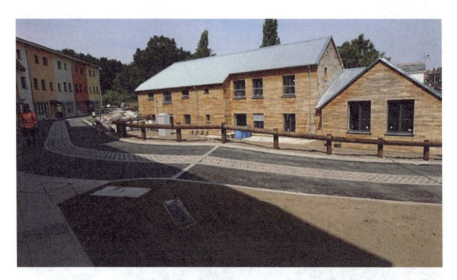

Fig. 5.10 Road entrance and flats (credit of Phil McGeevor)

Fig. 5.11 The old mill with new lift shaft entry (credit of Phil McGeevor)

5.2 The 50 Cases

Fig. 5.12 Rear of houses from ground level (credit of Phil McGeevor)

Fig. 5.13 Overview over the mill and the row houses (credit of Phil McGeevor)

5.2.4 *Casa Tucuna*[4]

Address: San Paolo, Brazil
  Setting: urban
  Status: established
  Size: 7 private dwellings
  Developer: Gabrielle Viseur

---

[4] Mari Pini.

Fig. 5.14 Internal view of a private apartment (credit of Mari Pini)

Architect: Eduardo Canals

Shared facilities and features: kitchen/dining room, meeting room, productive garden

Target: professionals

The experience of Casa Tucuna was born when the founder, Mari Pini, bought the current building and began to occupy it with different uses: at the beginning, placing the photo studio and sharing spaces with friends and relatives, and later, for 5 years, becoming a center for hosting students. Currently, the house is home to seven people, mainly professionals who are dedicated to artistic production.

The house, which has an area of 500 m^2, enjoys a great diversity of spaces: a large garage, garden, cellar, and mezzanine in addition to three bedrooms, a living room, dining room, terrace, and garden. The original rooms have been renovated to be adapted as a residence. Thus, on the ground floor of the house there are three separate and independent houses, with their own kitchen and bathroom: "Studio Mars" (which was originally a shed), "Grotta Greca" (the basement), and "Casa da Lua" (or the main house) where four people live, organized in three bedrooms and a mezzanine (Fig. 5.14). Casa Tucuna also has common areas: a patio (Figs. 5.15 and 5.16), an outdoor area for socializing (Fig. 5.17), a garden for growing vegetables, a kitchen and dining room for shared dinners and lunches, a living room that offers co-working opportunities and a laundry area. In addition to physical spaces, the community also looks for moments for meeting and action proposals by its members, such as meals, cultivation of the collective garden, body practices, and celebrations.

Casa Tucuna, which has not received any financial support from associations or public bodies, is nonetheless committed to the social impact, through the work of the founder Mari Pini, who, through the Public Design Institute, wants to promote

5.2 The 50 Cases

Fig. 5.15 The patio of the complex, over which the rooms face (credit of Mari Pini)

creative economy, improve public spaces, and promote conviviality. The plans for the near future are to create relationships with the next context involving neighborhood associations and turn Casa Tucuna into a residence for international researchers who want to live in a shared environment, rich in social relations, where they can develop projects related to art, design, and performance in the city.

5.2.5 Chiaravalle[5] [4]

Address: via Sant'Arialdo, Milan, Italy
    Setting: suburban
    Status: under construction

---

[5] Nadia Simonato (Cohousing.it).

Fig. 5.16 View over the green patio (credit of Mari Pini)

Fig. 5.17 The terrace of the complex, used as shared space for events (credit of Mari Pini)

Size: 50 families
Developer: Cascina Gerola
Architect: Bunch (concept)—Arch. Raul Bertolotti, Milan
Shared facilities and features: shared open green spaces, with productive garden, common rooms for shared activities
Target: mixed families
Website: www.cohousing.it/portfolio/cohousing-chiaravalle

5.2 The 50 Cases

Fig. 5.18 Render of an external view of the main complex, which is a restored historical cascina (credit of Newcoh srl—Cohousing Project Cohousing Chiaravalle)

Surrounded by the green agriculture fields of the South Agriculture Park of Milan, this project is located in a rural landscape rich in historical identity: the project began with the restoration of the historical buildings of Cascina Gerola (Figs. 5.18 and 5.19), a rural farm dating back to the seventeenth century, located in the vicinity of Chiaravalle Abbey, a building of the twelfth century, reference for the cultural and religious heritage of North Italy. Despite this rural settlement, the project is very well connected to the city of Milan, thanks to the metropolitan infrastructural system passing nearby. This project, which involves almost 50 families, came to life in the spring of 2016, after processes of design developed with the future residents, who represented the starting group of co-housers, enlarged by the arrival of new neighbors. The final project proposes productive gardens and orchards (25,000 m^2) in places previously occupied by green fields surrounding the farm. This is because the co-housers are not sharing just spaces or amenities (400 m^2), they are also promoting common activities and they are participating in common initiatives, like 0 km shopping, alternative transportation methodologies, etc. The eco-friendly approach of the project is also demonstrated by the use of contemporary sustainable building technologies and materials (Figs. 5.20 and 5.21). Due to the high level of social and environmental innovation, this project participated in Horizon 2020, on the topic of the reuse of existing buildings.

Fig. 5.19 Render of exterior spaces of the community (credit of Newcoh srl—Cohousing Project Cohousing Chiaravalle)

Fig. 5.20 Internal view of shared spaces in the main building (credit of Newcoh srl—Cohousing Project Cohousing Chiaravalle)

5.2 The 50 Cases

Fig. 5.21 Internal view of a private apartment (credit of Newcoh srl—Cohousing Project Cohousing Chiaravalle)

5.2.6 Coflats[6] [5]

Address: Bath St, Stroud (Gloucestershire), UK
  Setting: urban
  Status: established (2015)
  Size: 9 apartments
  Developer: Cohousing Company (David Michael)
  Architect: Peter Holmes
  Decision-making: consensus
  Shared facilities and features: kitchen, dining area, guest bedroom, laundry, gymnasium, sauna, workshops for crafts and woodwork
  Target: mixed families and individuals
  Website: www.coflats.net

Stroud is a town with a great propensity for co-housing: Coflats Sladbrook is the third co-housing community in this town. The philosophy of Coflats is inspired by the Isokon Building in Hampstead, a 1930s innovative community housing project where several artists, writers (such as Agatha Christie), some designers from the Bauhaus experiences, philosophers, etc. lived together, sharing a dining room and a laundry. With the same spirit, Coflats proposes a restored derelict building located in the town center (Fig. 5.22), where nine units (Fig. 5.23), with different features, have been realized following eco-friendly criteria. From every flat it is possible to reach the "common house" (Figs. 5.24 and 5.25) with shared spaces for everyday

---

[6] David Michael.

Fig. 5.22 External view of the building, from the street. The solar panels on the roof and facades are clearly visible (credit of David Michael)

Fig. 5.23 Kitchen in a private apartment (credit of David Michael)

5.2 The 50 Cases 137

Fig. 5.24 Common entrance in the building (credit of David Michael)

Fig. 5.25 Entrance and kitchen of the common building (credit of David Michael)

life. This aspect, together with the fact that some private storage rooms are located on the underground floor, allows the development of flats with lower dimensions. In the common house, residents can also share spaces for special activities, such as a gym, sauna, dining room, and a spare room for guests. In the patio garden and in the kitchen, co-housers regularly eat and relax together, but they also organize events opened up to the wider community. According to the co-housers' opinion, "sustainability is a clearly fixed aim, reached through a virtuous practice of restoration, use of building eco-techniques, car sharing and, mainly, through shared life."

5.2.7 CoHousing Israel (CHI)[7]

Status: planning

Size: 22 households

Target: Intentional Senior Community

CoHousing Israel (CHI) is still in the development process, but it seemed important to present this experience here, as it will be the first experience of contemporary co-housing in this country, where the practice of shared living is part of the culture. In fact, learning from this tradition, CHI aims to propose a community capable of offering a lifestyle alternative to the isolation in which today's older people, as well as other weaker groups, can often be found. The project is therefore presented as a place where people can grow old together in a communitarian atmosphere, emphasizing equality, inclusion, and environment. To do this, the project creates a community in which the co-housers are asked to be active in the operational management of the physical complex, the community culture and activity, and to support one another. The project involves the construction of a total area of 3300 m², including 30 private apartments varying from 45 to 95 m², in 3–6 storey buildings, and several shared indoor and outdoor spaces (350 m²). The 50 or so people who will join the community (now living in different parts of Israel) have signed a set of by-laws guiding the co-op's management, and will sign a member's agreement indicating the rules of living together, which the community members who actually move in together will create. Moreover, since the project refers to people in the 50–75-years age group, to buy into the co-op at least one member of the household must be within this range. Unlike kibbutzim (which were defined as communes, not cooperatives), the apartments are considered private property; there will be considerable common space for shared activities (kitchen/dining, laundry, garden, storage, exercise, tools, library, etc.); this also enables smaller apartments and savings in shared expenses. A relevant aspect of the architectural design is that it seeks to stimulate social interactions within the CHI community and between internal and external communities, encouraging activities with the neighbors such as tutoring, book clubs, volunteering, community gardens and other interactive enterprises.

---

[7] David Kurz.

5.2.8 Copper Lane [6, 7]

Address: Copper Ln, London N16 9NG, UK
   Setting: urban
   Status: established (2015)
   Size: 6 apartments
   Developer: self-development
   Architect: Henley Halebrown Rorrison Architects
   Decision-making: consensus
   Shared facilities and features: playroom, workshop, and laundry
   Target: mixed families
   Website: www.archdaily.com/580881/1-nil-6-copper-lane-n16-9nshenley-halebrown-rorrison-architects

Being the first case of co-housing reality in London, Copper Lane represents an innovation for the housing reality in the English capital. Copper Lane is a residential complex composed of six families, living in separate units, who did not know each other before starting this experience of co-housing. These families, with the support of the architectural studio Henley Halebrown Rorrison, contributed to designing the project, transforming in physical spaces the needs of each of them. The co-housers decided to gather the dwellings around a great common room where a playroom, a workshop and a laundry are shared (Fig. 5.26). The common room's roofing becomes the private open-air space for the families (Fig. 5.27). For the centrality of the common spaces, to keep all the necessary standards to preserve the indisputable privacy of each family, the dwellings have the main view over an external green ring, rather than facing inward toward the court. This green ring is a wide common

Fig. 5.26 Common area in the ground floor (credit of Ioana Marinescu and David Grandorge)

Fig. 5.27 Common terrace for shared activities (credit of Ioana Marinescu and David Grandorge)

garden for collective use, planned to emphasize the idea of sharing within the community (Figs. 5.28 and 5.29), to maximize the use of the external spaces and to recall a feeling of an intrinsic link with the environment, thanks to the local flora and fauna planned inside. The eco-friendly approach of the complex is also pursued by following the Passivhaus standards.

5.2.9 *CosyCoh*[8] *[8]*

Address: via Luigi Alamanni, Milan, Italy
  Setting: urban
  Status: established (2010)
  Size: 8 apartments
  Developer: Garden Estate srl + Cohousing.it promoters

---

[8] Nadia Simonato (cohousing.it).

5.2 The 50 Cases 141

Fig. 5.28 External spaces of copper lane (credit of Ioana Marinescu and David Grandorge)

Architect: Offarchitects
Decision-making: consensus
Shared facilities and features: common room for shared activities, laundry and open-air spaces (terrace)
Target: young/young families
Website: www.cohousing.it

Coming from the restoration of an existing building in Milan for handmade craft, this project is dedicated to young people and families up to 35 years old (Fig. 5.30). During the first stages of the design, the co-housers elaborated (1) the values and the rules, creating the bases of the community, and (2) the design of the shared spaces. CosyCoh, which is totally based on private investments, is the first rent co-housing project in Italy; this means that the residents are part of the community even if they do not own the apartments. Anyway, after they reach the age of 36, they can remain in the community by purchasing the dwelling at a favorable price (and showing the

Fig. 5.29 External spaces of copper lane (credit of Ioana Marinescu and David Grandorge)

potential creditors their capacity to sustain monthly payments). If any of the resident should leave, the new co-housers have to sign the manifesto of values and rules. In addition to these eight rented private housing units, this complex is composed of some shared spaces: a common room of 65 m^2 on the top floor, with an open-air terrace and a communal laundry underground (Figs. 5.31 and 5.32). The limited amount of shared spaces is related to the limited dimensions of the project itself, which actually represents an experiment for this kind of economical solution. Actually, according to the developers, the correct size for having a "minimum" critical mass is 30 units. Thanks to the design of the facades, the architects wanted to declare clearly the residential function of the building in a context that sill preserves some factory heritage aspects.

5.2 The 50 Cases 143

Fig. 5.30 View of the building from the street (credit of Newcoh srl—Cohousing Project Cohousing Cosycoh)

Fig. 5.31 Common area of the community for the bike parking (credit of Newcoh srl—Cohousing Project Cohousing Cosycoh)

Fig. 5.32 Wall for messages in the common area (credit of Newcoh srl—Cohousing Project Cohousing Cosycoh)

5.2.10 Cranberry Commons [9]

Address: 4272 Albert St, North Burnaby, V5C 2E8, Canada
  Setting: urban
  Status: established (2001)
  Size: 22 apartments
  Developer: self-developed with Cohousing Development Consulting
  Architect: Birmingham & Wood
  Decision-making: consensus
  Shared facilities and features: kitchen, dining area, lounge, children's area, guest room, meeting room, common laundry, teen room, workshop, courtyard and community gardens
  Target: intergenerational
  Website: www.cranberrycommons.ca
  This 22-home multifamily residential building can easily be considered an "urban village"; its great social interconnection, mixed with an efficient use of the resources, is the core concept of this community. Starting in 1998, a selected professional team provided services to a group of people who wanted to create a community that allowed for more connection with their neighbors, to manage the process of community creation: from the regulatory authorities' approvals to the building of membership with an organizational structure based on consensus decision-making. The future residents joined together to create a development corporation, which acted as developer of the project, and the community was completed in 2001 (Fig. 5.33). The dwellings (with dimensions varying from 45 m^2 to 120 m^2) and the

5.2 The 50 Cases 145

Fig. 5.33 External view, from the street, of Cranberry Commons (credit of Ronaye Matthew)

common house (315 m^2) are located around central spaces (Fig. 5.34), defining the outdoor areas with the community garden and courtyard (Fig. 5.35), which cover an underground parking area and storage. The shared facilities are the core of the design, with a kitchen and dining area (shared meals are organized at least once a week), lounge area, children's area, guest room, meeting room, common laundry, teen room, and workshop. A strong environmental sensitivity is shown with attention given to these communitarian issues, but also with care for the ecosystem. Sustainability represents one of the highest values of the community, which won the 2002 Environmental Award from the City of Burnaby.

5.2.11 Doyle Street[9] [10]

Address: 5514 Doyle St, Emeryville, California, 94608, USA
  Setting: urban
  Status: established (1992)
  Size: 12 apartments
  Architect: McCamant and Durrett
  Decision-making: consensus

---

[9] Evelyn Herrera.

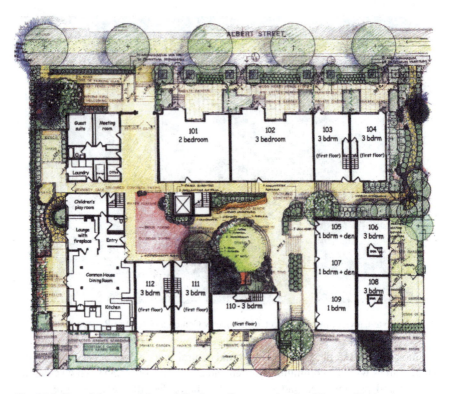

Fig. 5.34 Plan of the ground floor of Cranberry Commons (credit of Ronaye Matthew)

Shared facilities and features: kitchen for the shared dinner, a workshop, a laundry, a children's area, and a landscape patio

Target: families, artists

Website: www.emeryville-cohousing.org

Doyle Street has a long history and it is undoubtedly an international reference for the world of co-housing. Established in 1992 as one of the first co-housing projects in the USA, it was designed by two members of the community, Katy McCamant and Charles Durrett, who are also the founders of the US co-housing movement. The building, formerly a warehouse, has been restored and now the complex hosts several activities, such as small businesses, artist lofts, cafes and single-family homes, whose vivacity has been able to renew not only the spaces of the building, but also the spirit of the local community. The complex is composed of two buildings generating an "L" shape. Six units and the common house come from the renewal of the industrial building, while three townhouses are the new constructions. These residential loft-style units have dimensions of between 65 and 150 m^2 (Fig. 5.36). In the free area formed by the two blocks, a parking and the playing area

5.2 The 50 Cases 147

Fig. 5.35 Open-air distribution system, which becomes a common shared space for the community (credit of Ronaye Matthew)

Fig. 5.36 External view of some apartments of Doyle Street (credit of Evelyn Herrera)

Fig. 5.37 External view of common areas and private terrace in Doyle Street (credit of Evelyn Herrera)

Fig. 5.38 External view of the common area in the ground floor (credit of Evelyn Herrera)

are sited. The common spaces occupy an area of almost 200 m^2 with a kitchen for the shared dinner (two to five per week), a workshop, a laundry, a children's area, and a landscaped patio (Figs. 5.37 and 5.38). While the building underwent several changes in terms of spaces and functions, the co-housing organization proved to be very inclined to resist transformations and adapt to the changes.

5.2.12 Drivhuset [11]

Address: 24 Niels Ebbesen Vej, Randers, Denmark
  Setting: suburban
  Status: established (1984)
  Size: 18 apartments
  Architect: Niels Madsen
  Decision-making: consensus
  Shared facilities and features: central common space, with kitchen and services
  Target: families
  Website: www.coldhamandhartman.com/upload/documents/GreenMultiFamily.pdf

Drivhuset co-housing is the result of the renovation project of an old factory in Randers (Denmark) (Figs. 5.39 and 5.40). The complex is characterized by one large indoor atrium, bordered on both sides by two-storey buildings, which host the private dwellings (Fig. 5.41). In this greenhouse of 600 m^2, which works as the main common area, children can freely play in safety and adults can find opportunities to develop social relationships. Besides being the place where the idea of community becomes reality, this shared space characterizes the architectural feature of the project, including the fact that the name itself refers to this architectural element: in fact, Drivhuset literally means "greenhouse." Anyway, the communitarian street is not just a matter of common or iconic issues, but this element is used to temper the incoming fresh air and create a warm clime outside the dwellings. It is interesting to

Fig. 5.39 Entrance to Drivhuset co-housing (credit of Grace Kim)

Fig. 5.40 External view of the dwellings and their private gardens (credit of Grace Kim)

observe that, in terms of typology, the "communitarian street" also characterizes other co-housing projects: Wind Song in Canada, Àdalen in Denmark and Jernstøberiet, or, typologically similar, albeit in the open air, Swan's Market. In Drivushet, other common area, such as kitchen and living room, are located at the ends of the central corridor (Figs. 5.42 and 5.43).

5.2.13 Earthsong Eco-Neighbourhood[10] [12]

Address: 457 Swanson Road, Ranui, Auckland 0612, New Zealand
  Setting: semi-urban
  Status: established (2002)
  Size: 32 residences (65 residents)
  Developer: self-developed
  Architect: Bill Algie
  Decision-making: consensus
  Shared facilities and features: common house and common green, providing space and facilities for planned and informal activities
  Target: intergenerational, mixed
  Website: www.earthsong.org.nz

---

[10] Robin Allison.

5.2 The 50 Cases

Fig. 5.41 The internal distribution system, which becomes a relevant common area for informal meetings (credit of Grace Kim)

This project, which was the first experience of co-housing in New Zealand, is strongly based on the idea of an eco-friendly community; in fact, the aim was to value differences in age, and social and cultural backgrounds, to keep the community alive and rich in social life. Moreover, to strengthen the communitarian fabric, several activities are organized by the community: in particular, two formal dinners are organized every week, and an annual mid-winter festival brings the community together. To preserve the human-based design of the community, the cars are parked at the edge of the community, to leave the neighborhood peaceful and to free space for living; at the same time, to guarantee and facilitate walkable accessibility, the area is provided with paths with gentle gradients (Fig. 5.44). The community is flanked by an organization called "Earthsong Centre Trust" for environmental education and by a company called "Walk to Work Eco-Developments Limited," which has the aim of creating a green business cluster beside the community site. The community's buildings (Fig. 5.45) have also been

Fig. 5.42 External view of the common room of Drivhuset (credit of Grace Kim)

Fig. 5.43 Kitchen and dining room (credit of Grace Kim)

5.2 The 50 Cases

Fig. 5.44 Masterplan of Earthsong (credit of Robin Allison)

Fig. 5.45 External view of the common house (credit of Bill Algie)

Fig. 5.46 External view of a private dwelling in Earthsong (credit of Bill Algie)

constructed following eco-friendly principles, such as the use of nontoxic, low-energy materials, the development of permaculture gardens, the adoption of a rainwater catchment system, and the installation of solar water heating. Moreover, the shared garden contributes to 0 km production, providing seasonal fruit and vegetables for common use and to supplement common meals. While each private house and its own unit title are owned by single co-housers, the common parts are owned by all the households (Figs. 5.46 and 5.47).

5.2.14 Ecosol[11] [13]

Address: Faenza, Italy
  Setting: peri-urban
  Status: established (2012)
  Size: 13 families
  Developer: Group of co-housers Ecosol
  Architect: Arch. Luca Rigoni
  Decision-making: consensus
  Shared facilities and features: multifunctional room with kitchen and services, laundry and storage

---

[11] Luca Rigoni.

5.2 The 50 Cases

Fig. 5.47 External view of a private dwelling in Earthsong (credit of Bill Algie)

Target: mixed families
Website: www.ecosol-fidenza.it

Ecosol is a renowned project in the Italian panorama of co-housing, developed with the idea of becoming a replicable experience for the territory, where humans can again become the creators of the sustainable context for living. The project experience started in 2006, when several citizens, coming from different experiences of cooperation, sharing and environmental associations, began to meet to discuss the possibility of living in a friendly neighborhood based on social relations and eco-friendly approaches. The process for the definition of the community and the intervention took several years of discussions, with changes in group members and clarifications about the path to travel. From the meetings emerged the request for common spaces and services such as car sharing, babysitting, guesting, gardening, etc. At the end, the final group was composed of 13 diversified families, whose members all participated in the self-construction of the building. To control the costs, all the members participated in different building groups (technical elements, common spaces, social needs, etc.) on the construction site, which started in 2011 and ended in 2013, when the community started to inhabit the spaces (Figs. 5.48, 5.49, and 5.50). After the construction of the building, the community had to face some natural challenges coming from discords and divergences of opinion, but the social activities, previously planned, helped maintain a friendly environment. Bimonthly dinners, monthly meetings, shared works, and learning activities are at the core of those social moments, which often open the doors to the close neighborhood. Among all those relevant aspects, there are two particularly significant issues

Fig. 5.48 Front façade of Ecosol (credit of Luca Rigoni)

Fig. 5.49 Rear façade of Ecosol, with the pergola and the garden for common events (credit of Luca Rigoni)

Fig. 5.50 External distribution system (credit of Luca Rigoni)

that must be underlined. The first one is that the construction of an apartment was designed to host people in a state of "guided autonomy/situations in need of cordial accompaniment" up to the moment of self-sufficiency. The second one is related to the description of the community building process on the web page of the community, where the community explains how the technical problems met in reaching sustainable solutions (Fig. 5.51) can be easily solved if at the base of the process there is a social cohesion within the promoter group.

5.2.15 Emerson Commons [14]

Address: 1317 Parkview Drive, Crozet, VA 22932, USA
    Setting: rural
    Status: established (2019)
    Size: 26 units
    Developer: Peter Lazar Sheeflee
    Decision-making: consensus

Fig. 5.51 Picture taken during the construction showing the solution of straw bales for thermal insulation (credit of Luca Rigoni)

Shared facilities and features: shared indoor spaces include a common house with a kitchen and dining area, guest rooms and recreational spaces. Shared outdoor space include parking, walkways, open green space, a pool, an orchard, gardens, and a creek

Target: mixed families

Website: www.emersoncommons.org

Over a property of approximately 2.5 hectares, Peter Lazar, the current president of the Cohousing Association of the USA, decided to install the Emerson Commons Cohousing—a co-housing experience aimed at encouraging mutual assistance and association among the members of the community, to create a place for living, rich in social relations and environmentally friendly. Around a "turn-of-the-century farmhouse," restored after the purchase to establish the common house, 26 new private residences for the members of the community were built. These dwellings have five different sizes to standardize the offer while leaving opportunities for choice and customization. The design of houses and spaces is, though, in the direction of sustainable construction: 65% of the land is left as a green space, the houses have high standards of thermal insulation and PV solar installations are located over all the buildings, making Emerson Commons the first all-solar community in Virginia.

Before starting the shared-life experience, community members drew up a common vision and values charter, which all new members have to sign, allowing the community management to be supported in the future. Friendship, family, diversity, a balance of privacy and community, access to shared resources (common resources can also be used by single members for "private" events), a lower cost of living, a lower environmental impact and safety are the main issues around which the development and the management of the community revolve. On the same values chart, it is indicated how regular events are organized to manage shared spaces and activities and to enjoy the community life. This is because the activities take on a key role in helping to form a community, be they works for the maintenance of gardens and buildings, or cultural and fun activities. In particular, some relevant activities promoted by the community concern helping and caring for children and the elderly and a car pool.

5.2.16 Forgebank[12] [15]

Address: 9 Forgebank Walk, Halton, Lancaster, UK
  Setting: semi-urban
  Status: established (2012)
  Size: 41 families
  Developer: Lancaster Cohousing; self-developed
  Architect: Ecoarc
  Decision-making: consensus
  Shared facilities and features: common house with kitchen and dining/living area, guest rooms, children's room, laundry, food shop, workshops, outdoor spaces
  Target: intergenerational
  Website: www.lancastercohousing.org.uk
  Comprising 41 one- to three-bedroomed dwellings built to Passivhaus standards and common facilities, Lancaster Cohousing can be considered a neighborhood based on ecological values. The private and common spaces have been developed by the team of architects, together with the residents in a participatory design process. The main building follows the course of the river (Fig. 5.52) and a linear pedestrian street (Fig. 5.53), which connects the houses. The common house is in this row and includes a kitchen and dining/living area (for regular common dinners and communal activities) (Fig. 5.54). In front of this common house there are other shared facilities, including guest rooms, a kids' room and a food store. All the dwellings and the common house face the river, and have private decks and communal gardens (Figs. 5.55 and 5.56). These are not the only green spaces, since a common orchard and woodland are located within the area of the project. The community has also refurbished an old factory, Halton Mill, to create a work and events space for small businesses, artists and freelancers. It is interesting to underline the presence of this "co-working" space, linked with the co-housing, which renewed the historical coexistence of working and living spaces.

---

[12] Alison Cahn.

Fig. 5.52 View of Forgebank from the river (credit of Lancaster Cohousing)

Fig. 5.53 View of the community's street (credit of Lancaster Cohousing)

5.2 The 50 Cases 161

Fig. 5.54 Interiors of the common house (credit of Lancaster Cohousing)

Fig. 5.55 External spaces of dwellings and common house, facing the river (credit of Lancaster Cohousing)

Fig. 5.56 External spaces of dwellings and common house, facing the river (credit of Lancaster Cohousing)

5.2.17 Frog Song[13] [16]

Address: 101 Ross Street, Cotati, California, USA
  Setting: urban
  Status: established (2003)
  Size: 30 apartments
  Developer: Ross Developments, LLC
  Architect: McCamant and Durrett, The Cohousing Company
  Decision-making: consensus
  Shared facilities and features: common house with kitchen, multipurpose room, hot tub, truck, workshop, garden, children's room, guest rooms, laundry room, solar panels providing most of the electricity, and 21 electric vehicle charging stations
  Target: mixed families
  Website: www.cotaticohousing.org
  This intentional community of 30 households was formed by a group of individuals who started meeting and planning in the summer of 1998. Many of them had had previous co-housing experiences in California. Frog Song is not organized around any particular spiritual, philosophical, or ecological theme other than the universal co-housing principles of friendliness, social cohesion and connectedness, and sharing of resources. The aim of the community, as with most co-housing

---

[13] Peter Berking.

Fig. 5.57 Shared space outside the common house (credit of Peter Berking)

communities, is to create an environment that supports and sustains harmonious interpersonal relations and personal growth and fulfilment. It does this through a combination of design of physical spaces, sharing of personal and community-owned amenities, and member participation in community work activities. The core of the community is the common house (Fig. 5.57) located at one end of the site, which has shared facilities, including: a kitchen, a multipurpose room (which serves as a dining room as well as a musical performance space), a living room, a laundry room, guest rooms, a library, and a children's room. Next to the common house is a workshop with industrial power tools and a music rehearsal space above it; also the open-air spaces are design to encourage meetings (Fig. 5.58). Shared meals (dinners and Sunday breakfasts) are held 12–16 times a month. Member participation in work projects and tasks is part of Frog Song community life. Members meet once a month to manage the development and the business of the community. The government of the community is by consensus. Facilitators and conflict resolution procedures are used if needed. The 30 private dwellings are organized in 6 buildings with different dimensions: three with four units, two with five and one with eight. Each private unit has its own amenities, which are standard for most condominium units. Units are mostly owner-occupied and not rentals. The location of the community in a downtown area, with a public path going through the site, promotes interaction with neighbors in the town (Figs. 5.59 and 5.60). Within the site there are also 620 m^2 of commercial retail spaces owned by Frog Song, which was required during the development by the Municipality of Cotati. Rent income from these offsets monthly homeowner association fees.

Fig. 5.58 Open-air meeting point for common activities (credit of Peter Berking)

Fig. 5.59 Path and gardens which serve as distribution system among the dwellings (credit of Peter Berking)

Fig. 5.60 Path and gardens which serve as distribution system among the dwellings (credit of Peter Berking)

5.2.18 Il Mucchio[14] [17]

Address: Via Silvio Venturi, 7, Monte san Pietro (Bologna), Italy
  Setting: rural
  Status: established (2010)
  Size: 4 families
  Developer: self-development
  Architect: co-housers
  Decision-making: consensus
  Shared facilities and features: kitchen, workshops, laundry, garden, orchard, wood, pasture
  Target: families of friends
  Website: www.cohousingilmucchio.it

This study case can perhaps be considered one of the cases in which the alliance between humanity and environment is realized by the strength of the people's will. This co-housing was born on the hills (Fig. 5.61) nearby the northern Italian city of Bologna, starting from the life vision that eight young guys had almost 30 years ago. To realize their dream, this group of friends cooperated in a lot of activities: sharing time, money, decisions, energy and skills for several years, they became the developers and the builders of their own house. The complex occupies 600 m² of

---

[14] Michele Branchini.

Fig. 5.61 View from Il Mucchio, over the Po valley (credit of Michele Branchini)

Fig. 5.62 Buildings of Il Mucchio, restored form a former farm (credit of Michele Branchini)

rural houses and the fields of a small farmhouse called "il Mucchio" (Figs. 5.62 and 5.63). The intervention was focused both on the private houses and on the wide common spaces and services: from the kitchen, with the room for shared meals, to the workshops for hobbies and for DIY works. The 50,000 m^2 of green outside spaces allow a strong relation with the environment and nature, which provide fruits and vegetables thanks to productive gardens and orchards (Fig. 5.64). Inside the property there is also a wood, pasture, and a stable for animals. Small activities in the form of selling handicraft and agricultural production contribute to the maintenance of the little farm.

5.2 The 50 Cases 167

Fig. 5.63 The courtyard of Il Mucchio, after a party open to the local community (credit of Michele Branchini)

Fig. 5.64 Lavender field with the co-housers performing an open-air concert (credit of Michele Branchini)

5.2.19 Itaca[15] [18]

Address: via Faenza, Modena, Italy
  Setting: peri-urban
  Status: established (2010)
  Size: 12 families
  Architect: Emilia Amabile Costa
  Decision-making: consensus
  Shared facilities and features: living room with kitchen, guest room, workshop, garden, laundry
  Food: mixed families
  Website: www.antonellamonzoni.it/video.cfm

Itaca is the name of an Italian cooperative with indivisible property that has acted as the promoter and manager of this co-housing project. Itaca obtained from the Municipality of Modena the right to use the land for construction and from the Region Emilia Romagna some financing for the realization of the project. The ideological reasons behind this project are very strong: from the decision to have "indivisible property" as the key concept of the project to the eco-friendly techniques applied to the building (Fig. 5.65). Before starting this experience, not all the

Fig. 5.65 View of the main building of Itaca (credit of Maria Adelaide Frattin)

---

[15] Adelaide Frattin.

families knew each other, so the process to turn the first interested people into a group, capable of facing all the difficulties that a co-housing development requires had been a very strong and difficult effort promoted by the cooperative. Nevertheless, during the design and the construction phases, several families broke away and abandoned the process. The fact that the final residents followed the whole development, and in particular the design of the project, meant that every apartment was built following the desires of the inhabitants, so that everyone is different from everyone else. The project provides several shared facilities, both for technical services and for community creation, such as a living room with kitchen, a guest room, a laundry, a workshop and a garden, thought to increase sociability and improve the quality of life. At the moment, the co-housers feel that these shared facilities are underused in respect to the forecasts. According to the residents, the peaceful co-living environment depends on several clear rules for the use and management of the common spaces.

5.2.20 K1[16] [19]

Address: Graham Rd, Cambridge, UK
Setting: peri-urban
Status: under construction (2015)
Size: 38 apartments
Developer: self-commissioned
Architect: Mole
Decision-making: consensus
Shared facilities and features: kitchen, dining room, multipurpose room, playground for children and productive garden
Target: mixed
Website: www.cambridge-k1.co.uk
This project started in 2008, when the landowner and local authorities designed the land for a co-housing development and a number of people started to discuss the possibilities of living together in a co-housing community. So participatory processes for the design of the community started and, with the support of the architects, Mole, and the developers, town and Trivselhus, the project started to take shape (Fig. 5.66). The complex has 42 dwellings, most of which are along a pedestrian lane, which works as a common space for socializing activities (Figs. 5.67 and 5.68). At the core of the settlement, there is the common house, which hosts spaces for shared meals, and indoor spaces for gathering, playing and strengthening the sense of community. Moreover, a large common garden is thought to provide a safe space for the open-air activities of the children and for a small productive growing area (Fig. 5.69). The car parks are located on the edges of the plot, and there are 146

---

[16] Frances Wright.

Fig. 5.66 Model of one proposal for K1 co-housing used during the participatory design (credit of Frances Wright and Cambridge Cohousing Limited)

Fig. 5.67 Assonometric render of the complex of K1 co-housing (credit of Frances Wright and Cambridge Cohousing Limited)

5.2 The 50 Cases 171

Fig. 5.68 Ground floor of the complex, with gardens, private dwellings and common building (credit of Frances Wright and Cambridge Cohousing Limited)

cycling parking spaces within the complex. The private dwellings are composed of six different typologies and all, except one, have a private outside space or large balcony, underlining the core role of private property in co-housing projects. The management of the project is based on decision-making by consensus and a number of working groups. The community wants to underline that in the project there is no kind of activity based on a shared economy, while the main shared values are social and environmental sustainability.

5.2.21 Le Torri[17]

Address: Via delle Torri, 27/ 29/31, Florence, Italy
  Setting: urban
  Status: established (2019)
  Size: 7 families

---

[17] Anna Guerzoni.

Fig. 5.69 View of the external spaces and the common house (credit of Frances Wright and Cambridge Cohousing Limited)

Developer: self-development with public support
Architect: arch. Anna Guerzoni
Decision-making: Anna Guerzoni and the Association "Associazione Autorecupero Cohousing Le Torri"
Shared facilities and features: dining room, library, music room, workshops
Target: mixed families
Website: http://www.annaguerzoni.com/annaguerzoni/8-cohousing-casa-condivisa#!CA022

There are two aspects that, more than any other, led to the development of this project: the conception of housing as a natural right and the need to reestablish a social environment in contemporary cities, rejecting the idea that living should be confined to places of isolated life. The opening of a public tender, by the Tuscany region (2012), for the readjustment of publicly owned buildings through self-construction offered the opportunity for these ideas to materialize. To participate in the call for applications, an association had to be founded, interested residents found and the project prepared. The "Associazione Autorecupero Cohousing Le Torri" won this tender with this project, obtaining the opportunity to regenerate an old house, property of the municipality of Florence, and a non-repayable contribution for the purchase of building materials (Figs. 5.70 and 5.71). At the time of publication of the book, the recovery is in its final phase, but the association is

5.2 The 50 Cases 173

Fig. 5.70 The old building, which has been restored by the community to set the co-housing Le Torri (credit of Anna Guerzoni)

Fig. 5.71 Opening of the construction site (credit of Anna Guerzoni)

Fig. 5.72 Legend for the plans (credit of Anna Guerzoni)

1	APPARTAMENTO 1	78,31 mq
2	APPARTAMENTO 2	48,25 mq
3	APPARTAMENTO 3	71,84 mq
4	APPARTAMENTO 4	70,91 mq
5	APPARTAMENTO 5	79,15 mq
6	APPARTAMENTO 6	80,13 mq
7	APPARTAMENTO 7	69,83 mq
8 SPAZIO COLLETTIVO	APPARTAMENTO 8 PIANO TERRA 48,20 mq SEMINTERRATO 49,82 mq	98,02 mq

already active in promoting activities open to people in the neighborhood, involving other associations and city committees. In fact, together with the residential purpose, declined due to the typology of co-housing, the goal of bringing new vitality to the neighborhood through events and cultural activities open to citizens is very strong. The project involves the construction of eight apartments. Seven of these (about 60 m²) will be dedicated to the residence of families and the eighth (about 90 m² on two floors) will host common activities: dinners, children's games, events, etc. In particular, a basement with a vault is intended as a shared lounge where events can be hosted such as listening to music (Figs. 5.72, 5.73, and 5.74). Physically, the project was realized by the residents who, at weekends and during holidays, dedicated themselves to building their own houses (Fig. 5.75). In exchange for the regeneration of the building, which remains the property of the municipality of Florence, residents will have free housing for 30 years, after which the rent will be based on public building fees.

5.2.22 LILAC Grove [20]

Address: Victoria Park Ave, Kirkstall, Leeds, UK
    Setting: urban
    Status: established (2013)
    Size: 20 units
    Developer: partial grant given by Homes and Communities Agency
    Architect: White Design Associates
    Decision-making: consensus

5.2 The 50 Cases

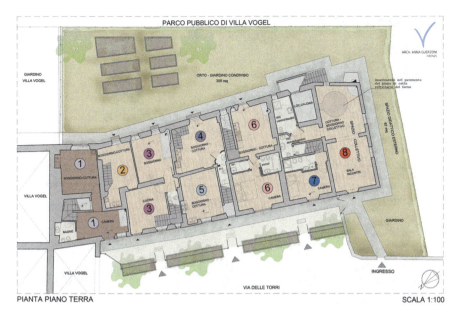

Fig. 5.73 Plan of the ground floor (credit of Anna Guerzoni)

Fig. 5.74 Plan of the first floor (credit of Anna Guerzoni)

Fig. 5.75 Member of the co-housing, working in the construction site to restore the existing building (credit of Anna Guerzoni)

Shared facilities and features: garden and common house with kitchen, meeting space, laundry, play area, office and guest rooms

Target: mixed families

Website: www.lilac.coop

LILAC (Low Impact Living Affordable Community) is considered a sustainable pioneering project and is the winner of several awards for its eco-based solutions. The complex, developed over an area previously occupied by a school, is composed of 8 houses and 12 flats (Fig. 5.76) organized around a common green space, which has the main purpose of reducing the ecological footprint of the community (Fig. 5.77). This courtyard arrangement is also planned to build community life, increasing relations between neighborhoods and intensifying the opportunities of socialization. For these reasons, the central green space is organized with a pond (also for sustainable urban drainage), some shared gardens and a playground (for kids and adults) (Fig. 5.78). The aim of the ecological feature of the project is fulfilled by the use of innovative building technologies and by an eco lifestyle that the co-housers, who sign a specific agreement concerning the community's philosophy at the beginning of their experience, have to follow. LILAC is open to the local community with some entrances (Fig. 5.79). The common house, the core of the community, is located near to the main pedestrian entrance. This position allows social interactions with the external local community, the hosting of occasional events, and permits the use of some facilities by external associations. To the co-housers, this common house offers a kitchen, a meeting space, laundry facilities, a play area, an office and guest rooms.

5.2 The 50 Cases

Fig. 5.76 LILAC from the balcony of a private apartment (credit of White Design)

Fig. 5.77 View of LILAC complex from the garden (credit of White Design)

Fig. 5.78 Shared the garden, with the small lake and the platform for shared activities (credit of White Design)

Fig. 5.79 A secondary entrance to the garden of LILAC (credit of White Design)

5.2.23 Los Portales[18] [21]

Address: Finca los portales, s/n 41230, Castilblanco de los arroyos, Seville, Spain
Setting: rural
Status: established (1984)
Size: 21 private dwellings and 12 guest dwellings
Developer: Gabrielle Viseur
Architect: Eduardo Canals
Decision-making: sociocracy, dreamwork as well as process work, forum tools, family and systemic constellations
Shared facilities and features: kitchen/dining room, meeting and games room, library, large hall for events and celebrations, workshops, swimming pool, bakery, agricultural hangars, cheese factory, livestock farm, oil mill, orchards and greenhouses
Target: eco aldea community
Website: www.losportales.net
In the late 1970s, various people (from different working and social backgrounds), who tried to change their way of life through Jungian psychology and dreamwork, decided to form a group with the aspiration to live in a community. Los Portales was born from this group, which in 1984 moved from Brussels to the countryside of Seville in a natural environment that had hardly been contaminated and was suitable for agriculture (Fig. 5.80). The goal was, and still is, to find a form of living that would be sustainable from a human and environmental point of view, in which to recover the connections and relationships between people and the natural environment (Fig. 5.81). The community is not conceived as an end in itself, but as a "means of creation" and "path of transformation." According to the members of the community, the close relationship with nature and the productive activity serve not only to rediscover a physical dimension where spiritual wellness can be reached, but also to relive, through the work, community relationships that have been lost in the contemporary individualistic society (Figs. 5.82, 5.83, and 5.84). For these reasons, the community has a strong relation with the environment, preventing erosion and desertification and working on the production of energy and resources with the desire to reach self-sufficiency. To date, wind, solar, and water systems have led to a self-sufficiency, up to 95%, of electricity. The controlled use of the natural resources of the hills (wood and water) and the production of olive oil, bread, goat cheese, wine, vegetables, grain, and hay for the animals and medicinal plants allow the community to have an extremely valid production of resources to cover the greatest needs of the people. The built environments of the community combine residential and productive spaces: a main building of 2500 m^2 is the core, with a kitchen, multifunctional room, library, workshop, 12 dwellings, 2 productive hangars, a small cheese factory and some restored buildings to be used as housing for members and visitors. The close relationship with nature and the compliance with

---

[18] Kevin Lluch.

Fig. 5.80 General view over Los Portales (credit of Kevin Lluch)

Fig. 5.81 Space for meeting and socialization in the landscape (credit of Kevin Lluch)

5.2 The 50 Cases

Fig. 5.82 Night social in the main courtyard (credit of Kevin Lluch)

ecology and permaculture also bring inspiration in the way the community manages its social and psychological dimensions, evolving over the years, to meet the changes at collective and global levels. The will to impact the wider community encourages the community to spread its experience, offering life experiences in Los Portales that can range from organizing regular guided visits, workshops and training to offering temporary short experiences and retreats.

5.2.24 Milagro Co-Housing[19] [22]

Address: 3057, N. Gaia Place, Tucson, Arizona, USA
  Setting: suburban
  Status: planning started in 1996|moved in 2002 and 2003
  Size: 28 homes
  Developer: community members
  Architect: Morton and Mackey
  Decision-making: consensus
  Shared facilities and features: common house with kitchen, meeting room, library, playroom and laundry; outdoor play areas, garden, swimming pool and community workshop

---

[19] Kenneth Schachter.

Fig. 5.83 Socialization under a pergola in a courtyard of the complex (credit of Kevin Lluch)

Target: intergenerational living
Website: www.milagrocohousing.org

The Milagro community sits on 43 acres in a residential neighborhood near the Tucson mountains, minutes away from downtown Tucson. Its members share a passion for community, the environment, and sustainability. The community's social life includes holidays and special celebrations, meetings, film nights and regular once-weekly community meals. Members meet monthly, except during a summer hiatus, to discuss and manage community business. Decisions are made by consensus at monthly community meetings with at least half of the homeowners present. Milagro's committees are responsible for the day-to-day management of

Fig. 5.84 A common party in the main courtyard of the complex (credit of Kevin Lluch)

the community: reporting back to members via written committee meeting minutes and seeking additional guidance when needed at monthly community meetings.

Milagro is designed to encourage daily interaction among its members. With the exception of two disability access homes, parking is remote from the homes and all of the front doors open onto a lush, wastewater-irrigated common area. A curving path leads homeowners through the common area from the ends of the community to the common house and the front gate at its center (Figs. 5.85 and 5.86). The common house provides space for meetings, gatherings, celebrations, family events, and meals. In addition to its large event space, the common house has a commercial-style kitchen, a library, a children's playroom, bathrooms, and a laundry room. Two outdoor children's play spaces, the pool, and a grassy entry circle surround the common house (Fig. 5.87). Milagro's strong ecological and environmental focus is evident in its passive solar-oriented adobe block construction, rooftop hot-water heating systems, wetlands wastewater treatment and subsurface irrigation system, the permaculture design of the interior common area with fruit-bearing trees, widespread homeowner adoption of rooftop photovoltaic solar panels, and the lifestyle choices of community members.

5.2.25 Moora Moora [23]

Address: 109 Moora Road, Mount Toolebewong, Healesville, Victoria, 3777, Australia

Fig. 5.85 Garden and a private house of Milagro co-housing (credit of Kenneth Schachter)

Fig. 5.86 The path, which distributes to the dwellings passing inside the park (credit of Kenneth Schachter)

5.2 The 50 Cases 185

Fig. 5.87 A playground for kids in the park of Milagro co-housing (credit of Kenneth Schachter)

Setting: pasture surrounded by native forest
Status: established (1970s)
Size: 30 units/houses in six clusters
Developer: self-developed
Architect: several over the years—some owner-built
Decision-making: aim for consensus, otherwise by majority vote
Shared facilities and features: several common buildings
Target: community-minded people—currently a mix of ages, singles and families
Website: www.mooramoora.org.au

An intentional cooperative community with a long history, since its history dates back to 1972, when seven young professionals formed a planning group. This group started the journey of Moora Moora, which has evolved a lot over the years, from its initial pioneering of a "progressive utopian community"—building of the community and regeneration of the land—to a more established community environment today. The community exists in a naturalistic mountain environment (Fig. 5.88), maintaining, however, a certain proximity to urban settlements and services (the city of Melbourne is 55 km from the community). Attention to social, educational, and ecological aspects is at the core of the first manifesto generated by the starting community, which also expresses the will to create a cooperative community, where diversity in personality and lifestyles can shape the social environment (Figs. 5.89 and 5.90). Members contribute different skills, and there is an emphasis on shared work, skill sharing, and an educational role in the wider community. The whole community is off-grid, and both individual members and the cooperative are involved in solar power initiatives in the wider area. The community is organized

Fig. 5.88 View over Moora Moora co-housing and the Australian landscape (credit of Joanne Broughton)

Fig. 5.89 Structural straw bales used for thermal insulation (credit of Joanne Broughton)

into six clusters located on the 245-hectare property, all within ten minutes' walk from the main common building (Fig. 5.91). To preserve the natural environment and to facilitate social interactions, each cluster occupies half a hectare. These clusters consist of groups of four to six dwellings and some shared infrastructure, while the main central lodge and learning center hosts the shared facilities for community

Fig. 5.90 Labyrinth stone in Moora Moora (credit of Joanne Broughton)

Fig. 5.91 Some of the private and common buildings in the first phases of the construction (credit of Joanne Broughton)

Fig. 5.92 The build of Mount Camphill for co-housing (credit of Vicky Sime)

space, workshops, and events. While the whole community initially planned the shape of the clusters, houses are planned and built by individual members.

5.2.26 Mount Camphill[20] [24]

Address: Faircrouch Lane, Wadhurst, East Sussex, UK
 Setting: rural
 Status: established (2017)
 Size: 3 houses providing 16 supported co-housing rooms, 8 unsupported co-housing rooms and 4 apartments
 Developer: The Mount Camphill Community
 Architect: Tangent Space
 Decision-making: consensus
 Shared facilities and features: common houses with kitchen, living room, dining room, bathrooms, laundry, gardens and workshops
 Target: care and support for young people
 Website: www.mountcamphill.org
 The Mount Camphill Cohousing Community is an intentional community—it is the adult program of The Mount Camphill Community (a specialist college dedicated since 1970 to the education of students with learning disabilities) (Fig. 5.92). The learning activities of the college are very varied, including many cultural events with an accent on craft, art, therapies, and caring for the land. This mix of activities makes life in the community very diverse and active, just like the co-housing community. Precisely for this reason, alongside the central nucleus of the college, the community decided to pioneer a cohousing project, open to anyone, including people with and without a learning disability. It aims to enhance all aspects of an individual's life, promoting a cultural community that is socially supportive and

---

[20]Vicky Sime.

5.2 The 50 Cases

Fig. 5.93 The management of the complex is carried out directly by the members of Mount Camphill Community, aiming to enhance all the aspects of individual life. (credit of Vicky Sime)

inclusive. The fact that the Mount Camphill Community is focused on people with learning disabilities gives the co-housing project an opportunity to experiment with new forms of living that separate housing from care and support. Fundamentally a wish to community-build together is at the heart of the impulse and anthroposophical care and support are offered if requested.

The Mount Camphill Community shows the potential for real change in the way that we can live and community-build for the future. As a social therapeutic answer to life sharing, co-housing aims to promote the well-being and the integral development of each person (with or without a learning disability) in its relationship with the social and natural environment: integration between private and public spheres; opportunities for sharing experiences and lifelong learning; enhancement of the principles of respect, sharing and responsibility toward others; awareness of not being alone, but of being part of a community. In particular, it should be noted that to facilitate the achievement of all these goals, all members are encouraged to participate, according to their abilities, in the care of the whole co-housing community, including the land, social enterprise workshops and in the management of the community (Figs. 5.93, 5.94, and 5.95). Currently, the project is expanding to accommodate six more supported co-housers; they are actively fundraising to expand and renovate their third community house in the local village of Wadhurst.

Fig. 5.94 Maintenance activity within the community (credit of Vicky Sime)

Fig. 5.95 Handcraft product from the activity of the community (credit of Vicky Sime)

5.2.27 Munksøgård [25]

Address: Munksøgård 4000 Roskilde, Denmark
  Setting: peri-urban
  Status: established (2000)
  Size: 5 communities of 20 apartments each

5.2 The 50 Cases

Decision-making: 1 board for each community

Shared facilities and features: 1 main common house with main services and 5 common houses for each community

Target: mixed families, cooperative young, and older people

Website: www.munksoegaard.dk

Munksøgård is one of the most renowned co-housing communities in the international panorama. To support the concept of "diversity," this main community is composed of five different groups: (1) single families' owners of the dwellings (Fig. 5.96); (2) members of a cooperative association and common owners (Fig. 5.97); (3) young people, on rent (Fig. 5.98); (4) older people, on rent (Fig. 5.99); and (5) every age, on rent. The buildings of the last three communities are owned by the Roskilde Building Association, which rents the dwellings. These five communities are set in five groups of 20 terraced houses, arranged in a horseshoe shape around an old farm, located at the core of the settlement and with all the facilities shared among the main community, such as a cafe, a vegetable and gift shop, a co-working and bicycle repair workshop and storages. At the same time, each one of the five communities has its own common house for all its own "restricted" activities, such as meetings, shared meals, etc. Because of the dimensions of the community, the decision-making processes are quite complex, with five boards managing the five communities and one board managing the commonly owned properties and

Fig. 5.96 The Ejer community of condominium ownership in Munksoegaard co-housing (credit of Grace Kim)

Fig. 5.97 The buildings of the rental cooperative community for families (credit of Grace Kim)

their activities. Moreover, there are several working groups in which all the cohousing members are supposed to participate—some with technical, others with social and integrating aims. The communities are characterized by an extensive recycling that reaches levels of 90–95%, and also by a sustainable approach, as can be appreciated due to the construction techniques.

5.2.28 *Mura San Carlo [26]*

Address: via Samoggia e Seminario, La Mura San Carlo, San Lazzaro di Savena, Bologna, Italy
    Setting: semi-urban
    Status: completed (2014)
    Size: 12 families
    Developer: E'cohousing and Cooperativa Cohousing Mura San Carlo
    Architect: tamassociati
    Shared facilities and features: cinema, multipurpose room, kitchen, music room, workshop and services
    Target: mixed families
    Website: www.tamassociati.org/PAGES/SLZ/SLZ_cohousing.html

5.2 The 50 Cases 193

Fig. 5.98 The buildings of the rental cooperative community for young people (credit of Grace Kim)

Fig. 5.99 The buildings of the elder people community (credit of Grace Kim)

Fig. 5.100 Front façade of Mura San Carlo co-housing (credit of Andrea Avezzù)

Fig. 5.101 Mura San Carlo from the surrounding sport fields (credit of Andrea Avezzù)

The project of Mura San Carlo (Figs. 5.100 and 5.101) has been developed by the association "È/co-housing," born in 2009 from the ideas of some friends interested in the living issues related to space sharing and ecology. The association contributed to spreading the concept of co-housing in Italy and realized three projects. In the realization of this specific project, È/co-housing received the support of the architectural studio TAMassociati, which contributed as a promoter, manager, and

5.2 The 50 Cases

Fig. 5.102 Night view of the building (credit of Andrea Avezzù)

facilitator of the whole process. The design process was very important, since the main idea of the association was to achieve strong environmental and economic sustainability, combined with features of social innovations, such as participatory design, personalized private spaces and elective neighborhood. Four years of meetings brought the definition of the final design, which offers 12 dwellings and common spaces (cinema, laundry, multipurpose room with kitchen, music room, and workshop). The design also proposes common services, such as car sharing, a time bank and an ethical purchasing group, which allow an approach of respect for common goods and for sharing dynamics to be developed. The common spaces are realized on the ground floor to facilitate the integration among the co-housers and the social relations with the local external community, often hosted for educational-didactic activities (Figs. 5.102 and 5.103).

5.2.29 Nubanusit[21] [27]

Address: 7 Callies Cmn, Peterborough, New Hampshire, USA
 Setting: rural
 Status: established (2007)
 Size: 29 apartments
 Developer: Nubi River Partners; co-founders living in the community
 Architect: O'Neil Pennoyer Architects, Groton
 Decision-making: consensus
 Shared facilities and features: farm, natural land and common house with kitchen, living room, playroom, guest rooms, office

---

[21] Clive Russ.

Fig. 5.103 Insertion of Mura San Carlo in the rural landscape (credit of Andrea Avezzù)

Fig. 5.104 Overview of Nubanusit co-housing (credit of Clive Russ)

Target: mixed families
Website: www.peterboroughcohousing.org

This project, even if it is totally immersed in a natural environment (Figs. 5.104 and 5.105), is just 90 min away from the city of Boston. This location allows the co-housers who live here to have different lifestyles: from the professional point of view, some of them work from home, some commute and some are retired; similarly, children in the community attend public, private or home school. A mission ("To work together to nurture and support each other and the land on which we live") and a list of values, based on respect, environmental stewardship, openness and interdependence, help this community to live in harmony and to avoid or, in some cases, to solve problems. Sustainability is one of the main features desired by the community and, for this reason, several design issues have been considered: each home is super-insulated; natural and non-toxic materials are used and the heating is provided by a hot-water collector and a central wood pellet-fired heat plant. The common house is the core of the community and it is designed to facilitate fortuitous meetings and to strengthen interpersonal relationships. This building includes spaces for socializing, such as a living room, a kitchen, a kids' play space, and a guest room (Fig. 5.106). While each private house has its own yard, the com-

Fig. 5.105 View of the snow-covered complex (credit of Clive Russ)

Fig. 5.106 Shared dinner in the common building (credit of Clive Russ)

munity can count on 113 acres, shared, with a farmland, fields, and woodlands (Fig. 5.107). The farm, developed on 5 acres, supports a Community Supported Agriculture (CSA) vegetable farm, in whose program most of the residents participate.

Fig. 5.107 The houses of Nubanusit stand out from the crop fields (credit of Clive Russ)

5.2.30 OWCH [28]

Address: Union St, Barnet, London, UK
  Setting: urban
  Status: established (2017)
  Size: almost 20 apartments
  Developer: Hanover Housing Group
  Architect: Pollard Thomas Edwards (PTE)
  Decision-making: consensus
  Shared facilities and features: common living room with kitchen and services, garden, workshop, and guest room
  Target: women older than 50
  Website: www.owch.org.uk

Almost 20 women, between 50 and 80 years old, coming from different life experiences and having decided not to live alone, composed the starting group of this co-housing project. They decided to plan together what would have become the first co-housing for older people in the UK. The idea comes from the desire to offer and receive mutual support, living in an environment capable of encouraging exchanges and social relations. After several years of study on the benefits of co-housing and construction issues, the starting group received initial help to materialize the idea from the Hanover Housing Group. The architectural studio Pollard Thomas Edwards designed, with the OWCH members, a sustainable proposal respectful of the urban context and with internal flexible spaces, capable of responding to the changing needs of older people. The community was established in a

Section
From street to garden

New Ground CoHousing, High Barnet

Fig. 5.108 Section of the building on the main street, highlighting the common and private spaces (credit of Pollard Thomas Edwards)

purpose-built block in North London, with a morphology that evokes the one of the demolished buildings (Fig. 5.108). The new constructions, with an L shape (Fig. 5.109), create a central space, used as a shared garden (Figs. 5.110, 5.111, and 5.112). The community wants to become a social resource for the neighbors and an example for other older people. For this reason, monthly, all-day meetings are organized inside the community to explain this reality to external associations and groups. This project is recognized as one of the most impactful and well-known projects of co-housing in the world, winning nine awards between 2016 and 2017, including the European Collaborative Housing Award 2017.

5.2.31 Pacific Gardens Co-housing Community[22] [29]

Address: 347 Seventh St., Nanaimo, B.C., Canada
  Setting: urban
  Status: established (2009)
  Size: 25 residences (almost 54 members)
  Developer: the 13 original residents and shareholders
  Architects: Jolyon Brown and Ian Niamath
  Decision-making: consensus

---

[22] Kathryn Hazel.

Fig. 5.109 Site plan of the old women co-housing (credit of Pollard Thomas Edwards)

Fig. 5.110 View of the dwellings from the shared garden (credit of Pollard Thomas Edwards)

5.2 The 50 Cases 201

Fig. 5.111 Private apartments and common rooms facing the central courtyard (credit of Pollard Thomas Edwards)

Fig. 5.112 View of the courtyard from one of the balconies of the complex (credit of Pollard Thomas Edwards)

Shared facilities and features: kitchen, dining room, office, children's playroom, arts and crafts room, exercise room, music room, and guest rooms

Target: multigenerational

Website: www.pacificgardens.ca

The adventures of this community started in 1998 when the Land Committee was established and proceeded with the choice of the land that 9 years later saw the beginning of construction. The site, with an area of 1.76 hectares, is former farmland, whose property is bordered by the Chase River and has a wonderful view of Mt Benson. Nestled in a forest, it also has a pond with ducks and full-throated tree frogs. The building (Fig. 5.113), designed to maximize the green areas and social interaction, offers 25 units with different dimensions opening onto the glass-covered atrium, which leads to the common house, creating an indoor street and meeting area for residents (Fig. 5.114). This shared space occupies three levels with a walkout patio and an upper deck. The main level has the kitchen, dining room and meeting room (which leads to the patio with a nice panorama), a music room and a children's playroom. The lower level has a workshop, exercise room, recycling and bicycle storage room, while the upper one has two guest rooms and a bathroom, a conversation lounge, a sewing room, and an arts and crafts room. Some 740 m^2 are occupied by these indoor common spaces and by some outdoor common spaces: organic gardens (Fig. 5.115), a children's play area, a fire pit, a pond, a river, and a

Fig. 5.113 Exterior of the common house and private dwellings (credit of Alina Abbott and Pacific Gardens)

Fig. 5.114 Internal courtyard, which works as common area and entrance to private houses (credit of Alina Abbott and Pacific Gardens)

forest. These common spaces are the scene of several shared activities: one pot-luck dinner every week; two dinner clubs that meet once a week; afternoon tea once a week; a coffee hour once a week; and resident parties once a month. For this project the city of Nanaimo gave an award of excellence to architects Jolyon Brown and Ian Niamath in 2010.

5.2.32 Pioneer Valley [30]

Address: Pulpit Hill Road, North Amherst, Massachusetts, USA
  Setting: rural
  Status: formed (1989), established (1994)
  Size: 32 apartments
  Developer: self-developed
  Architect: Kraus-Fitch Architects
  decision-making: consensus
  Shared facilities and features: kitchen/dining room, children's playroom, guest rooms, laundry facilities, library, sauna, exercise room
  Target: mixed families

Fig. 5.115 Productive garden cultivated in front of the building (credit of Alina Abbott and Pacific Gardens)

Website: www.cohousing.com

This community, built in 1994 over 9 hectares, is considered the first co-housing neighborhood in the eastern United States. Its system of values is very strong and based on diversity—as Pioneer Valley co-housing a is multigenerational project, with members coming from several different experiences—and on the desire to support communitarian life, providing interactions and communications that go over the physical boundaries of the community, involving external residents in several activities and associations' memberships. The design of the project focuses on these points, so favors pedestrian movements, relegating the car parks to the edges (Fig. 5.116). The residential development is composed of 32 households and five rented units of energy-efficient housing, connected by paths, open fields and gardens (with a hen house and a labyrinth) (Fig. 5.117). At the entrance to the plot is the main building: a 420 m^2 common house, with an annexed workshop, a home-office building and numerous other play structures. The common house hosts several shared spaces, including, among others, a large dining room and supporting commercial-grade kitchen, two guest rooms, a children's play room, laundry facilities, meeting and living rooms, and a library, but also an on-site home office with shared equipment. The common house, considered an extension of the homes, is the central space where all the common activities are performed, starting from the two common dinners per week, with an average attendance of three-quarters of the community's

Fig. 5.116 Entrance of the foot path of pioneer Valley co-housing (credit of John Fabel)

members. Moreover, the celebration of various holidays and events becomes a regular activity. The renowned sustainable approach is guaranteed thanks to the orientation (Fig. 5.118) and the super-insulation of the buildings, but also thanks to the partial self-production of food from the common garden.

5.2.33 Pomali [31]

Address: Landersdorf 108, 3124 Wölbling, Austria
  Setting: rural
  Status: established (1970s), organization founded (2003) and co-housing implemented (2013)
  Size: 30 units (29 flats with different dimensions, one tiny house)
  Developer: mainly Martin and Petra Kirchner
  Decision-making: circular structure, in consent, sociocratic
  Shared facilities and features: kitchen, bistrot, living room, children's room, meditation room, spa, permacultural gardens, playground, workshop, shared office, car sharing, food co-op, chicken farm
  Target: mixed families
  Website: www.pomali.at

Fig. 5.117 The complex of private housing forming the community (credit of John Fabel)

Fig. 5.118 Sunset over Pioneer Valley co-housing (credit of John Fabel)

5.2 The 50 Cases

Pomali, which literally means "slowly," is the acronym from the German words for "practically," "ecologically," "connectedly," "attentively," "passionately," and "inclusively" (*Praktisch, Oekologisch, Miteinander, Achtsam, Lustvoll* and *Integrativ*), which represent the way in which the residents want to live. In particular, the goal of the community is to cultivate positive relationships with each other and live as ecologically, sustainably, and diversely as possible (Amali Dicketmüller). This community is composed of 77 individuals of different ages (0–78), activities and backgrounds (families, elders, single parents, young couples, singles), who contributed to founding the community in 2009 or joined the project in the following years. The wealth of experiences and the generational difference are aspects deliberately sought, to allow all the members of the community to grow and so be able to grow new communitarian forms of social, environmental, and economic development. In this perspective, the diversity that co-housers can have is very respected, in particular with regard to spirituality and politics, and moments of confrontation are enhanced to generate new ideas on how to live the relationship with the environment. The co-housing project is developed in a condominium with 29 private residential units (from 50 m^2 to 120 m^2, with great attention to the practice of passive construction) (Fig. 5.119), large common areas and 10,000 m^2 of green spaces (designed with the idea of an "edible landscape," created by orchards and vegetable gardens in permaculture plantation) (Fig. 5.120). The common areas are mainly concentrated in the community building, in the center of the residential complex. Here the main activities of the community take place, from eating together to playing, celebrating and discussing, but also the meetings with the neighbors of the wider community take place here (Fig. 5.121). A lot of social activities are thought to connect the individuals with the environment and the community, such as heart-sharing circles, celebrations, rituals, coaching, yoga, singing, and dancing

Fig. 5.119 Private dwellings and common garden (credit of Amali Dicketmüller)

Fig. 5.120 Communitarian work in the shared garden (credit of Amali Dicketmüller)

Fig. 5.121 Activity of the community in the common building (credit of Amali Dicketmüller)

(Fig. 5.122). Among the relevant aspects of sharing, the community offers a car-sharing service, composed of seven cars, and "shared shopping" (food co-op), in which the members buy and store all that the community needs and is not produced in the gardens.

Fig. 5.122 Winter ceremony in the garden (credit of Amali Dicketmüller)

5.2.34 Quattropassi [32]

Address: via Ponte Canale, Villorba (TV), Italy
   Setting: semi-urban
   Status: built (2015)
   Size: 8 houses
   Developer: self-developed with Cooperativa Pace e Sviluppo
   Architect: tamassociati
   Decision-making: consensus
   Shared facilities and features: wide open green space with a productive garden and a central common house
   Target: families
   Website: www.tamassociati.org/PAGES/VLB/VLB_cohousing.html

Committed to ecological design, co-housing Quattro Passi became a resilient neighborhood, where the sharing philosophy is combined with environmental sustainability approaches. These approaches work on the different scales of intervention: on the urban scale the idea of "eco-neighborhood" wants to minimize the impact on the environment; on the local scale, the co-housing solution generates social relations; on the domestic scale, the "eco-houses" prioritize the attention to consumption of resources. The project started in 2009, promoted by the cooperative Pace e Sviluppo, with the design developed by Tamassociati with the participation of the future inhabitants, who have been key figures in defining the project. The first part of the meetings was about reflections, debates and generation of a concept, while in the second part, through activities of participatory design, the designers defined spaces, services, common areas, and dwellings. These phases of planning between designers and people have been fundamental in creating a place capable of

Fig. 5.123 View of the complex from the main street (credit of TAMassociati)

Fig. 5.124 The central shared garden and the private houses lead to the common house, core of the complex (credit of TAMassociati)

responding to the people's needs and dreams. The result is this residential complex, composed of eight houses, one for each family making up the community, bounded by a wide green field that leads to the common building of 230 m², which is the central element of the intervention (Figs. 5.123, 5.124, and 5.125). Here, all the main common spaces are located: a multifunctional room, with a kitchen, a workshop for craft activities, a guest room, and the storage for the Solidarity Purchase

5.2 The 50 Cases

Fig. 5.125 External pedestrian path which leads to the private houses (credit of TAMassociati)

Groups (Fig. 5.126). The management of all these spaces and of the common activities of the complex is in the hands of the co-housers.

5.2.35 Quayside Village[23] [33]

Address: 510 Chesterfield Ave, North Vancouver, British Columbia, Canada
 Setting: urban
 Status: built (1998) and established
 Size: 5 townhouses, 14 apartments, 1 commercial unit
 Developer: Quayside Village Cohousing group
 Architect: Richard Vallee
 Decision-making: consensus
 Shared facilities and features: dining room, kitchen, guest room, laundry room, office, domed meditation room, garden courtyard, and third-floor deck
 Target: mixed families
 Website: https://sites.google.com/site/quaysidevillage/home

---

[23] Kathy McGrenera.

Fig. 5.126 View of the complex from the common house porch (credit of TAMassociati)

    The community, already established, moved into the Quayside Village building in the summer of 1998. Its urban setting in the neighborhood of Lower Lonsdale provides a stunning panoramic view over the skyline of Vancouver and the natural surroundings of North Shore, Burrard Inlet and the Strait of Georgia (Fig. 5.127). Quayside Village is multigenerational and hosts couples, singles and families with every kind of belief, religion, or lifestyle. These differences among the community's members are reflected in the spaces, which include 19 residential units of different sizes and typologies: bedsits, flats (1, 2, and 3 bedrooms) and townhouses (2 and 3 bedrooms). These private dwellings and the common spaces are gathered in a dense area of almost 1000 m^2. The design establishes a central common courtyard that connects to all but one of the entrances to the private homes and creates a safe and welcoming environment to strengthen the social bonds of the community (Fig. 5.128). The common house of 230 m^2, a rooftop deck and a meditation room are shared among the community's members. Quayside Village was built on a site that formerly occupied three single-family houses, and the hardwood flooring, antique doors and stained-glass windows were reused for the new constructions (Figs. 5.129 and 5.130). A close-to-zero waste reuse and recycling program, outdoor laundry lines, extensive composting and food growing, and an initial greywater system are some of the sustainable strategies adopted by the community. The project has won several awards over the years for its social sustainability and environmental initiatives.

Fig. 5.127 Quayside Village seen from the street (credit of Grace Kim)

5.2.36 Radiance Cohousing[24] [34]

Address: 475 Avenue L South, Saskatoon, Saskatchewan, Canada
  Setting: urban
  Status: established
  Size: 8 units
  Developer: Radiance Cohousing Development Company
  Architectural design: BLDG Studio, with consultants' support
  Decision-making: consensus
  Shared facilities and features: common house, office area, guest room, garden, patio, car share, solar panels
  Target: mixed
  Website: www.radiancecohousing.ca

Radiance Cohousing is located in a central urban area of Saskatoon, bordering residential and industrial areas (Fig. 5.131). It is near a park, the river, public services, and downtown. The aim of the community is to create a multigenerational

---

[24] Shannon Dyck.

Fig. 5.128 The internal patio of Quayside Village (credit of Grace Kim)

environment that leads to a high quality of life for its residents and the surrounding community. The two main values that guide Radiance Cohousing are respect for people and environmental sustainability. The housing complex was designed by the residents with the support of BLDG Studio and built collaboratively with the aid of Renew Development Cooperative. Construction financing was provided by New Community Credit Union. The final result is a project composed of eight privately owned houses (Fig. 5.132), of different sizes and designs, built to achieve Passivhaus standards, with the support of Bright Buildings. Radiance Cohousing also partnered with the Saskatoon Carshare Cooperative to provide a shared electric car, as well as the SES Solar Coop to install a large solar photovoltaic array. Outdoor areas include individual and shared gardens, fruit trees and edible plants, an outdoor cooking area and patio, and water-conserving features (such as a 9500-L rainwater cistern and a dry stream). The landscape was designed by Green Edge Studio following the principles of permaculture design, and steps were taken to remediate the soil as the project is located on a former brownfield site. The common house—equipped with

Fig. 5.129 The front façade of the building (credit of Quayside Village Cohousing)

a full kitchen (Fig. 5.133), eating area, guest room, bathroom, office area, piano, and balcony—is accessed by residents and the community at large for events, meetings, celebrations, and learning opportunities. The basic idea of living alongside a community is that relationships are strengthened thanks to collaboration, distribution of leadership, shared events, spending meaningful time together, and sharing skills (Fig. 5.134). Moreover, the most important decisions for the management of activities are reached through consensus decision-making, which improves communication, listening and understanding, as well as leading to shared outcomes and decisions.

5.2.37 Rancho La Salud Village[25] [35]

Address: Carretera Poniente Chapala-Jocotepec 1259, Ajijic, Lake Chapala, Jalisco, Mexico
  Setting: semi-urban at lakeside
  Status: established (2012), some facilities in progress
  Size: 12 individual homes, 18 town homes plus 6 apartments in 3 acres
  Developer: Multiversity Corporacion SA de CV
  Architect: Rick Colishaw & Denia Navarro

---

[25] Denia Navarro.

Fig. 5.130 The front façade of the building (credit of Grace Kim)

Fig. 5.131 The snow-covered community seen from the street (credit of Shannon Dyck)

5.2 The 50 Cases

Fig. 5.132 Private apartment (credit of Shannon Dyck)

Fig. 5.133 Common house (credit of Shannon Dyck)

Fig. 5.134 Potluck with radiance cohousing members (credit of Shannon Dyck)

Decision-making: sociocracy via a self-governance body

Shared facilities and features: organic orchard, common house, spa and pool, multipurpose conference center, passive and active solar energy, and overall permaculture philosophy

Target: mixed

Website: www.rancholasaludvillage.com

In 2012, during the first stages of creating Rancho La Salud Village, the future co-housers started meeting and talking about what kind of community they wanted to live in: what core values they would be dedicated to and what principles would guide them along the way to set and manage the community. This co-housing is set in the center of Mexico, on the north shore of Lake Chapala, near the "magic" town of Ajijic, characterized by being a destination for retired foreigners. Actually, this co-housing community also refers to this kind of foreign public, proposing a lifestyle based on the principles of healthy living and lifelong learning. The project is based on a 400 m^2 central common house (a restored existing building) (Figs. 5.135 and 5.136), with several external shared facilities, including a conference center, a swimming pool (Fig. 5.137), a jacuzzi, an organic vegetable garden, and orchards. The layout is designed organically to promote human interaction and social capital. The garden houses, town homes (Fig. 5.138) and apartments surround these common facilities. The buildings' design focuses on the ecological sustainability of the project, with solutions proposed such as: passive solar design; use of solar panels to cover the total demand for electricity and hot water; recycling, collection and biodigester systems for water. Some teams and committees are formed to better follow

Fig. 5.135 View of the common house (credit of Jaime Navarro Miramontes)

Fig. 5.136 The patio of the common house (credit of Jaime Navarro Miramontes)

the needs of the community, but decision-making is always by consent. The main difference with this co-housing consists in the hired staff supporting the management of some provided services like pool maintenance, electricity, plumbing, etc., allowing the members to focus on the human system dynamics design (the "software")

Fig. 5.137 View of the common garden and some buildings from the common pool (credit of Jaime Navarro Miramontes)

Fig. 5.138 A private house (credit of Jaime Navarro Miramontes)

growing and awakening to higher levels of consciousness in the individual and the collective realms. This co-housing introduces new conversational technologies drawn from Non-Violent Communication (NVC) guidelines and integral living practices. The core values are centered in community, sustainability, multiversity and longevity.

5.2.38 Rocky Hill[26] [36]

Address: Black Birch Trail, Northampton, Massachusetts, USA
Setting: forest
Status: established (2006)
Size: 28 residences (almost 75 members: 50 adults and 25 children)
Developer: Doug Kohl, Kohl Construction
Architect: Coldham & Hartman Architects
Decision-making: consensus
Shared facilities and features: community garden, swings for children and sand-pit, sledding hill, 27 acres in total (9 occupied and the rest wooded, with trails)
Target: mixed families
Website: www.rockyhillcohousing.org
The concept of Rocky Hill is to create an intergenerational collaborative neighborhood located in the middle of the Connecticut River's Pioneer Valley, capable of recapturing the essence of community, and creating a human environment. The manifesto shows, in nine points, how the main goal of the project is the individual and the social relations that the community can build around them. For this aim, the physical design is considered fundamental in the enhancement of individual well-being, the sense of community and the harmonious relation with the environment. Its realization took 5 years from the participatory design to the construction. The project expands over a site of 3 hectares and there are 13 duplexes and two single-family homes (Fig. 5.139), while no future expansion is allowed by agreement with city's Conservation Commission. To facilitate the social relations, these dwellings are clustered in a figure of eight around the common house and all houses have a small front porch facing out into the center of the community (Fig. 5.140). The common house has a large multipurpose room, a family room, a guest room, two small sitting rooms, two bathrooms, and a full basement with ping pong, billiards, air ice hockey, sauna and a crafts area plus a washer and dryer (Figs. 5.141 and 5.142). Moreover, to get together the members of the community there are some groups (book club, garden club, film night, Actively Growing Older Gracefully, etc.), as well as some committees to manage the community (common house c., community life c., sustainability and environment c.). In Rocky Hill the attention given to sustainability and the environment is also remarkable, achieved as it is through energy-efficient construction, solar panels, recycling, composting and food gardening.

---

[26] Thomas RC Hartman.

Fig. 5.139 Masterplan of rocky hill co-housing, with the common house in the center (credit of Coldham & Hartman Architects)

Fig. 5.140 Pedestrian path among the community (credit of Richard Getler)

5.2 The 50 Cases

Fig. 5.141 Living room in the common house (credit of Richard Getler)

Fig. 5.142 Living room and kitchen in the common house (credit of Richard Getler)

5.2.39 Solidaria: San Giorgio[27] [37]

Address: via Ravenna 228, Ferrara, Italy
  Setting: semi-urban
  Status: built (2015)
  Size: 7 apartments
  Developer: Cooperativa Cohousing Solidaria
  Architect: Rizoma architecture
  Decision-making: consensus
  Shared facilities and features: multipurpose common room with kitchen, library and laundry
  Target: families
  Website: www.cohousingsolidaria.org

The first co-housing project in the Italian city of Ferrara appeared in the neighborhood of San Giorgio, which combines historical and naturalistic richness and is recognized as a heritage site by UNESCO. The first meeting, which defined the starting group, composed by seven families interested in the idea of co-housing, happened in 2008 and, some months later, the association Cohousing Solidaria, which promotes lifestyles based on sharing, respect and solidarity, was founded. From this moment, a long process of community building and definition of physical space, where to locate the community, started. According to the residents, these processes for the creation of the community were the most difficult, but important, ones. Once the community was established and the location fixed, a process of participatory design started in 2013 thanks to the support of Rizoma Architecture and the financing of an ethic cooperative bank (Banca Etica). In 2015, the project was completed (Fig. 5.143), and the community could settle in this complex of

Fig. 5.143 Rear façade and back garden (credit of Alida Nepa)

---

[27] Alida Nepa.

Fig. 5.144 Shared moment among the community (credit of Alida Nepa)

3500 m^2 (820 m^2 of residences) with at ground level the entrances for the private dwellings and a multipurpose room (50 m^2) with a kitchen, library, and laundry. Very significant are the relation of the community with the river which flows on the backside of Solidaria and the activities that the community performs here (Figs. 5.144, 5.145, and 5.146). Moreover, car sharing, productive gardens, ethical purchasing groups, and a time bank are shared among the residents. The complex won the international Green Building Solutions Awards 2015 for its environmentally friendly strategies.

5.2.40 Springhill[28] [38]

Address: Stroud, England, UK
 Setting: urban
 Status: established (2003)
 Size: 35 (20 houses and 15 apartments)
 Developer: Cohousing Company Ltd (David Michael)
 Architect: Architype
 Decision-making: consensus

---

[28] David Michael.

Fig. 5.145 Common activity in the behind canal (credit of Alida Nepa)

Fig. 5.146 Common space on the back side of Solidaria co-housing (credit of Alida Nepa)

5.2 The 50 Cases

Fig. 5.147 Main pedestrian path among the community, which works as distribution system and meeting space (credit of David Michael)

Shared facilities and features: open-air spaces and common house, with kitchen, multipurpose room, dining room, and children's room

Target: mixed families and individuals

Website: www.springhillcohousing.com

Established in 2003, this project was the first project developed by Cohousing Company Ltd, created and founded by David Michael, the managing director who bought the site in Stroud in 2000. He made the master plan, employed architects, arranged funding and recruited future residents. After that year, other residents joined the company and the project started with the support of Architype architects, who designed the layout and detailed design of each building. This was done with the participation of all the members of the community. After residents moved in to their houses and flats, they became equal directors of the company that owned the site. The residents are in charge of the management of the community and share common activities like cooking, eating together, childcare, and gardening. According to the architects, the result is a "remarkable sense of environmental and resident satisfaction, which can be attributed to the residents being part of the design and delivery of their own homes and having ownership over their design and construction." The concept of the community is based on: pedestrianization and a child-friendly site (Figs. 5.147 and 5.148); small private gardens contiguous to the

Fig. 5.148 Snow-covered main street, showing the different levels on which the community is set (credit of David Michael)

dwellings (Fig. 5.149); the common house as the core of the community (Figs. 5.150 and 5.151); and eco-sustainability as the focal point.

5.2.41 Stolplyckan [39]

Address: Föreningsgatan 35–59, Linköping, Sweden
  Setting: urban
  Status: established (1980)
  Size: 184 apartments in 13 buildings
  Developer: group of women with the municipality of Linköping
  Decision-making: committee elected by the community
  Shared facilities and features: 2000 m^2 of "living room": common kitchen, dining rooms, senior center, playground, workshops, guest room
  Target: families and social support
  Website: www.stolplyckan.nu
  Stolplyckan was born from the idea of a group of women who wanted to create a social environment in which to live and let the children grow. The idea of this first

5.2 The 50 Cases

Fig. 5.149 A private house (credit of David Michael)

Fig. 5.150 Dining room of the common house (credit of David Michael)

Fig. 5.151 Living room of the common house (credit of David Michael)

group of women was discussed and finally accepted by the local administration, which decided to realize a residential co-housing project, characterized by several shared spaces and facilities, in an area of Linköping Municipality, which was supposed to be intervened with an urban restoration. Today, Stolplyckan, thanks to its 13 buildings, with 184 apartments, linked by a system of paths and common spaces, represents the biggest reality of co-housing in the whole of Sweden (Fig. 5.152). Here, while 140 apartments are rented to residents that pay a moderate monthly fee to the municipality, the other 44 are allocated to older people. Totally aligned with the North European welfare approach, the local municipality promotes and supports this community. Moreover, thanks to the relation between the public and private sectors, many of the shared spaces of the community that during the day would remain unused are managed by the municipality until 6 pm for welfare activities (schooling support, elderly center, etc.). Stolplyckan adds to these facilities several other common spaces, including a sports center, a music room, workshops for wood and ceramics, and living and play rooms. The management of all these facilities, spaces and activities is directed by a local council. To increase the idea of community, every Monday one of the buildings prepares a common dinner for all the 13-building community. Additionally, the co-housers also write produce a newspaper with the life events of the community.

Fig. 5.152 Aerial view of Stolplyckan (credit of Pål-Nils Nilsson, licensed under the Creative Commons Attribution 2.5; https://io.wikipedia.org/wiki/Arkivo:Linköping_-_KMB_-_16001000 013234.jpg)

5.2.42 Sunflower[29] [40]

Address: Champagne-Mouton, France
  Setting: rural
  Status: established and developing (property purchased 2013)
  Size: 10 units
  Developer: self-developed
  Architect: self-designed
  Decision-making: consensus
  Shared facilities and features: communal house, courtyard, gardens, vegetable allotment, field
  Target: mixed families
  Website: www.sunflowercohousing.org.uk

Sunflower Cohousing is a community developed in a small rural settlement in the Nouvelle-Aquitaine region, but although in a rural location, the community has a very good infrastructural connection. This community, open to persons of different ages, abilities and nationalities, is structured around the goal to create, in the long term, a sustainable environment where living is based on mutual support. This

---

[29] Martin Prosser.

Fig. 5.153 View of the courtyard defined by the residential building and garages (credit of Martin Prosser)

support is considered at different levels within the co-housing community, where life is organized around work, rest and play in a balanced relation between privacy and communitarian solidarity.

To realize this dream of living in co-housing, the founder group formed a Civil Real Estate Company, SCI (Société Civile Immobilière) du Tournesol, which is a nonprofit-making company, with the aim of buying land for the community. The choice landed on a plot of 40,000 m^2 with an old farmhouse (Figs. 5.153 and 5.154), which became, after restoration (Fig. 5.155), the core of the community thanks to its c-shape with attached barns that define the main courtyard. An orchard and a vegetable garden have been formed within a part of the remainder of the plot (Fig. 5.156). The community has detailed planning permission for five residential units, two of which are under construction, with a further outline permission to extend the community to another five units, all facing the central courtyard and with a private garden to the rear. The Civil Real Estate Company SCI du Tournesol also supports the management of the community and owns community land, buildings and equipment. Residents can be part of the co-housing either by way of simple participation with the co-housing group or by being stakeholders of the real-estate company. In this last case, all those new members who match the investment of the existing stakeholders will also have the same rights in terms of capital spending decisions. In any case, all the residents have the same rights in respect of policy discussion and the same duties in terms of community work to support the management of the co-housing. The approach to environmental sustainability includes the preservation of water and energy and through a more responsible management of the locally available natural resources.

5.2 The 50 Cases 233

Fig. 5.154 View over agricultural hangar and the landscape (credit of Martin Prosser)

Fig. 5.155 The main residential building under restoration (credit of Martin Prosser)

Fig. 5.156 The shared productive garden (credit of Martin Prosser)

5.2.43 Swan's Market [41]

Address: Oakland, California, CA, 94607, USA
　Setting: urban
　Status: established (2000)
　Size: 20 apartments
　Developer: East Bay Asian Local Development Co. (EBALDC) (nonprofit)
　Architect: Michael Pyatok & Assoc., Peter Waller
　Decision-making: consensus
　Shared facilities and features: Swan's way, garden and a common house with kitchen and multipurpose room
　Target: mixed families
　Website: www.swansway.com

5.2 The 50 Cases

Fig. 5.157 Street view of Swan's Market complex (credit of Luca Boveri)

The area where this project is situated was previously occupied by the Oakland Free Market, which located here in 1917. From the 1950s the neighborhood of Oakland had been in decline, leading to the market (now called Swan's Market) to be closed in 1984. Among all the proposals presented to the municipality for renewing the area, the selected result combined the idea to preserve the historical heritage and to create a mixed use to reconnect the fragments of downtown Oakland and to create a friendly street life (Figs. 5.157 and 5.158). The mixed use came to include a children's art museum, dwellings, retail shops, restaurants and offices, creating almost 150 new jobs. In the general process of renewal, the co-housing group exerted, with its vision for community living, a powerful force in the development of the project. This renewal represents a really successful housing project because it embodies the combination among sustainable practices, such as (1) the preserving of historical heritage, (2) the use of urban mixed uses, (3) the reinforcement of the connection with the neighborhood and the social fabric, and (4) the creation of human-scale atmosphere (Figs. 5.159, 5.160, and 5.161). The project for Swan's Marketplace was completed recently but the new life brought here by the co-housing project is already promoting new initiatives every day.

Fig. 5.158 Exterior view of the complex that constantly remarks its identity related to the former market (credit of Luca Boveri)

5.2.44 Temescal Creek [42]

Address: Oakland, California, 94609, USA
  Setting: urban
  Status: established (1999)
  Size: 11 families
  Developer: self-developer
  Architect: The Cohousing Company/McCamant and Durrett
  Decision-making: consensus
  Shared facilities and features: kitchen and dining room, a sitting area, laundry, bathroom and kids' room
  Target: mixed families
  Website: www.cohousing.org/cm/article/temescal

5.2 The 50 Cases

Fig. 5.159 Public spaces outside the community (credit of Luca Boveri)

Fig. 5.160 Public spaces outside the community (credit of Luca Boveri)

Fig. 5.161 Public spaces outside the community (credit of Luca Boveri)

This co-housing project is interesting because of the procedure that led the community to buy and build the complex. The project started in 1999, after only three months of meetings among the residents, without having planned and fixed all the details of the project. For this reason, the complex was built in several different steps. Firstly, three adjacent 1920s duplexes were bought, then two houses from the same block, and thus, addition by addition, the community reached its final shape. It is interesting to underline that at the beginning there was not a common house, so for the first year, the private houses were the places in which, in rotation, the community met twice a week. Afterwards, for 3 years, an office, annexed to a new purchased house, was used as a common house, until the community could build the 83 m^2 official common house, including the kitchen and dining room, a sitting area, laundry, bathroom, and kids' room (Figs. 5.162 and 5.163). All the private dwellings are connected by a path which runs along the central areas (Figs. 5.164 and 5.165). This process is fascinating because it underlines a feature of all the co-housing projects, that is, being composed by two main layers: the communitarian

5.2 The 50 Cases 239

Fig. 5.162 Internal picture of the common house (credit of McCamant & Durrett Architects)

Fig. 5.163 Living room in the common house (credit of McCamant & Durrett Architects)

Fig. 5.164 Common garden and a private dwelling (credit of McCamant & Durrett Architects)

Fig. 5.165 Common garden and a private dwelling (credit of McCamant & Durrett Architects)

spirit and the physical spaces. As this project demonstrates, the spaces are important for facilitating and promoting the social relations, but a co-housing project cannot be considered such if the spirit is missing: the willingness is the real condition *sine qua non* that makes co-housing different from other ways of living. Temescal Creek can be defined as a "retrofit" community because of the idea of using an existing neighborhood to improve it and transform it, using mostly the existing housing.

5.2.45   Terracielo[30] [43]

Address: via Giusti, 28, Rodano (MI), Italy
  Setting: rural
  Status: built (2011)
  Size: 57 apartments
  Developer: Novum Comum
  Architect: Maiocchi Pellegrini Patergnani Architetti Associati
  Decision-making: –
  Shared facilities and features: multifunctional room, fitness area, sauna, guest house, laboratory, music room
  Target: families, couples, singles
  Website: www.terracielo.biz

In the Italian panorama, the building development promoted by a private company venture is typical of the area of Milan, where co-housing is seen as an economic opportunity for developers. Promoted by Novum Comum and its partners, this complex is based on the idea that the economy of scale allows the quality of the living experience to be raised, and the construction costs controlled. In fact, this project has several shared facilities with high-quality levels, which are not common in other co-housing projects: in the central courtyard, surrounded by the private dwellings (Figs. 5.166 and 5.167), are placed the sauna, a soundproofed music area, a fitness area, a guest house and a children's playing area, but also spaces for an ethic purchasing group (Fig. 5.168). These shared facilities have been designed through a process of participatory design, which involved the future inhabitants. In the central courtyard, in addition to these facilities, are some gardens; some of them are shared among the whole community, some are held by the owners of ground floor apartments. The dwellings are offered with different dimensions and distributions, but all of them are characterized by a high-quality level of detail. As is understandable from the description and from the pictures, this project represents an approach to the practice of co-housing that is substantially different from several others presented in the book.

---

[30] Vittorio Travella.

Fig. 5.166 View of the complex TerraCielo from the roof (credit of Trivella s.r.l. with arch. Aldo Maiocchi-Ivan Patergnani-Antonio Pellegrini)

Fig. 5.167 The private houses and gardens define the courtyard with the common house (credit of Trivella s.r.l. with arch. Aldo Maiocchi-Ivan Patergnani-Antonio Pellegrini)

Fig. 5.168 Entrance to the common house and outdoor living space (credit of Trivella s.r.l. with arch. Aldo Maiocchi-Ivan Patergnani-Antonio Pellegrini)

5.2.46 Urban Village Bovisa[31]

Address: v. Maddalena Donadoni, Milan, Italy
  Setting: urban
  Status: established (2009)
  Size: 32 apartments
  Developer: Loftland (construction)—Cohousing.it (participatory design, community building, marketing)
  Architect: Luca Beverina & Ass.
  Decision-making: co-housers and manager
  Shared facilities and features: common kitchen, hobby room, open swimming pool, laundry room and common garden
  Target: young people, families, young couples and old people
  Website: www.cohousing.it/portfolio/urban-village-bovisa
  Nowadays, Bovisa is one of the most interesting areas of Milan, where its great industrial urban heritage has been regenerated thanks to numerous renewal projects developed here in the last few years. This co-housing project fits perfectly within this environment, representing at the same time (1) the cultural ability of building

---

[31] Nadia Simonato.

Fig. 5.169 Internal view of a private apartment (credit of Newcoh srl—Cohousing Project Urban Village Bovisa)

regeneration and (2) the local spirit of innovation, with this project being one of the first cohousing experiences in Italy. Urban Village Bovisa, in fact, was developed in 18 months, between 2007 and 2009, and now hosts 32 families in a structure that once was a cork factory; besides the private spaces (Fig. 5.169), the project arranges 200 m^2 of covered shared spaces and 400 m^2 of open-air shared spaces (Figs. 5.170 and 5.171), and a swimming pool on the terrace. The project is the result of a collaboration between the Politecnico di Milano and an agency for social innovation, which proposed a survey to understand the level of interest in the idea of co-housing in Milan. In 2006, the founder group, which now represents half of the actual residents, started the process of "community development": project managers and social facilitators (Cohousing.it) helped in reaching agreements, aiding the co-housers in the shared design phase. During these phases, which were developed through workshops and brainstorming sessions, the project managers used a "toolkit," composed of particular post-it and activity cards, to facilitate the dialog between co-housers and designers (Fig. 5.172). This work, which today is a reference among the practices of participatory design, has been useful for creating a sense of identity and defining the values that the community is based on.

5.2 The 50 Cases 245

Fig. 5.170 Courtyard of Urban Village Bovisa (credit of Newcoh srl—Cohousing Project Urban Village Bovisa)

Fig. 5.171 Entrance to the complex of Urban Village Bovisa (credit of Newcoh srl—Cohousing Project Urban Village Bovisa)

Fig. 5.172 Initial phase of participatory design (credit of Newcoh srl—Cohousing Project Urban Village Bovisa)

5.2.47 Vaubandistrict [44]

Address: Vauban, Freiburg im Breisgau, Germany
  Setting: urban
  Status: established (2004)
  Size: 5000 habitants
  Developer: Forum Vauban
  Architect: several
  Decision-making: several methodologies have been developed
  Shared facilities and features: common houses, sports facilities, activities, associations, etc.
  Target: wide-ranging
  Website: www.vauban.de

This project is on a scale totally different from the previous projects, even if the whole project cannot be considered a "co-housing" project due to several features that this intervention has. Vauban is a district of the German city of Freiburg; it is an urban intervention developed on land originally settled as a military base. After the area was abandoned by the armies, several associations, such as the "Forum Vauban," disputed with the municipality to ensure that the area would be developed in a contemporarily sustainable way. After several years, Forum Vauban proposed to the City Council its car-free and eco-friendly master plan, and finally this district was

5.2 The 50 Cases   247

Fig. 5.173 Solar panels in Vauban district (credit of Arnold Plesse, licensed under the Creative Commons Attribution 3.0; https://commons.wikimedia.org/wiki/File:Freiburg_071ss.jpg)

designed and built from 1998 to 2004, becoming an internationally renewed reference for its sustainable solutions. This urban project is presented in the book because almost 40% of the buildings have been self-developed by co-housing communities[32] and because the whole management of the project shows a communitarian and environmentally friendly based approach (Fig. 5.173). While several dwellings are occupied as co-housing, the municipality kept some dwellings as dormitories for the students of the local university. Several associations are working inside Vauban; in particular, they are involved in the fields of culture, kids, teens, neighborhood and ecology. Vauban is also a remarkable place because here several proposals of self-government have been experimented with, arriving at some proposals for solutions for social innovation.

5.2.48 Wandelmeent[33] [45]

Address: Wandelmeent 17, 1218 CN Hilversum, the Netherlands
  Setting: urban
  Status: established (1976)
  Size: 50 houses
  Developer: Stichting Woningcorporatie Het Gooi

---

[32] https://www.construction21.org/city/h/vauban-ecodistrict-freiburg.html.
[33] Judith Kortland

Fig. 5.174 Pedestrian path and open-air meeting space (credit of Judith Kortland)

Architect: L. de Jonge en P. Weeda
Decision-making: democratic voting
Shared facilities and features: common house, kitchen, living room, guest rooms, bar, fitness room, library, workshop, laundry, games room
Target: mixed families
Website: www.wandelmeent.nl

Wandalmeent is a very well-known reference in the world of co-housing. It was promoted in 1969 by a group of citizens and supported by the Dutch Ministry for Culture, Free Time and Social Solidarity. From the beginning, three main points were agreed upon as the basis of the project: (1) it would be available to every income bracket; (2) the initial number of apartments was fixed at 50; (3) the decision on how to manage the community would be made by the residents. The community is organized in ten clusters with five apartments and a common room for each. The space connecting these clusters is an open-air street, where the social activities of the community members can be performed in a safe environment (Figs. 5.174 and 5.175). Another building, the common one, houses a small kitchen and a living room for the main meetings, while other common buildings are dedicated to the free time of the co-housers. Every day, at 11 a.m. someone walks through the street and rings a bell to invite all the members to join for a cup of coffee together. The decisions regarding the management of the community and its activities are taken by every present co-houser during monthly meetings.

Fig. 5.175 Stairs to the common terrace (credit of Judith Kortland)

5.2.49 Windsong[34] [46]

Address: 20543—96th Avenue, Langley, British Columbia, Canada
　Setting: suburban (Greater Vancouver)
　Status: established (1996)
　Size: 34 town homes connected in one building
　Developer: Windsong Cohousing Construction Cooperative partnered with Northmark Projects
　Architect: Davidson, Yuen Simpson (DYS)
　Decision-making: consensus of owner members

---

[34] Howard Staples.

Fig. 5.176 The central green house, which works as distribution system and space for meeting and shared activities (credit of Miriam Evers)

Shared facilities and features: large common house, creek area, etc.
Target: families (with some home schooling)
Website: www.windsong.bc.ca

This co-housing experience is a very interesting project for the glass-covered street and the well-preserved natural environment: just 40 km east of Vancouver, this project proposes a combination of urban and rural lifestyle. The design scheme consists in 34 town homes arranged in two rows connected by a covered pedestrian street, which becomes an important element of common life (Fig. 5.176). Along this main axis, at the middle of the length, there is the common house, which, as a hinge, changes the direction of the street, organizing the houses—a street system with two main wings. The common street, which all the private units face, is covered by a greenhouse-style glass roof that allows for extra common spaces protected by the weather and the climate benefits of greenhouses to be achieved. The common house, which has a surface of 460 m^2, includes a generous common kitchen, a dining room (where shared meals are organized twice a week), a fireside area, a playroom, a multipurpose room, some workshops, an office and a guest room (Figs. 5.177 and 5.178). About half of the dwellings have a private yard, while the open shared space, managed by the community, includes productive gardens, playgrounds for children, relaxation areas and 4 acres in a natural state, crossed by the Yorkson Creek (Fig. 5.179).

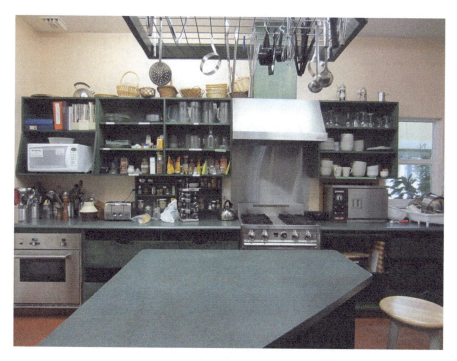

Fig. 5.177 The shared kitchen in the common area (credit of Miriam Evers)

Fig. 5.178 The inner playground (credit of Miriam Evers)

Fig. 5.179 The shared productive garden and the open-air playground (credit of Miriam Evers)

5.2.50 Wolf Willow [47]

Address: 530 Avenue J South, Saskatoon, Saskatchewan, Canada S7M2A8
  Setting: urban
  Status: group formed in 2008; moved in September 2012
  Size: currently 30 members in 21 units
  Developer: self-developed with Cohousing Development Consulting
  Architect: Mobius Architects, Sechelt, BC with RBM Architects, Saskatoon, SK
  Decision-making: generally open discussion, although not always full-consensus decision-making; formal vote on financial matters
  Shared facilities and features: kitchen, dining room, lounge, workshop, craft and music rooms, guest suite, laundry, sauna and exercise room, indoor car park, gardens (flowers, vegetables and fruit), rooftop decks
  Target: older adults without live-in children (residents range in age from 52 to 85)
  Website: www.wolfwillowcohousing.ca
  The meetings of Wolf Willow community started in 2008 from a fortuitous encounter between two distinguished groups with the desire to live in co-housing. With consensus decision-making, this merged group continued the meetings on a monthly basis for 18 months. After several crucial decisions on direction, e.g., single, not multigenerational, condominium legal form, not co-op, just seven people

5.2 The 50 Cases 253

Fig. 5.180 Wolf willow community from the street (credit of Louise Clarke)

were left. Nevertheless, this core group proceeded to buy a building site and engage a co-housing development consultant. Over the following 2 years almost 30 people were recruited who directly participated in designing the community's building and self-management policies designed to enhance friendship, cooperation, community, and environmental sustainability. In 2012, Wolf Willow became the first co-housing project in Saskatchewan and the first seniors' co-housing project in Canada (Fig. 5.180).

The site and structure were designed for aging in place and to be as "green" as possible within the technology available at the time and budget constraints: extra thick insulation in the exterior walls, high-efficiency boilers for heating, no air conditioning, composting and recycling facilities, no grass, community and individual garden plots, and a conduit for solar retrofitting (a solar installation came online in July 2019[35]) (Fig. 5.181). There is a total of 21 private dwellings (seven units on each of three floors) with wide, single-loaded corridors to optimize light and congenial sitting areas (Fig. 5.182). The common facilities (400 m²) (Fig. 5.183), built around a courtyard and gardens at street level, are linked by an elevator to the private units. The community activities include shared meals and celebrations, house

---

[35] Words of the co-houser Louise Clarke.

Fig. 5.181 Shared garden (credit of Louise Clarke)

Fig. 5.182 Distribution system to the private dwellings (credit of Louise Clarke)

5.3 Comments on the Cases

Fig. 5.183 Dining room and living room of the common area (credit of Louise Clarke)

concerts, welcoming friends and families to guest suites, and tasks associated with maintaining the building (voluntary, not prescribed). All residents are members of a governing council that meets ten times a year and participation remains gratifyingly high.

5.3 Comments on the Cases

The desire to present so many projects is that each reader can form a personal idea about the vast panorama of co-housing in the contemporary world. However, there are some aspects that I would like to highlight to express a couple of personal observations that I consider particularly relevant.

The first aspect that struck me in these years, moving into the world of co-housing, studying numerous cases and discussing with members of various communities, is the attitude that communities have toward the surrounding context and their intention to really pose as renewal vectors for the society. The decision to live in a community is not only dictated by the search for a better life but also by the desire to be part of a wider transformation. Sometimes communities have to close themselves to the numerous requests of people who want to take part in projects because the dimensions impose a limited number of members; however, the average attitude of the communities is absolutely not that of a closure. The opening that the communities offer is unprecedented in contemporary cities: although defining a circumscribed and protected space, communities paradoxically open up toward the city and society. In numerous case descriptions, for example, one can read how numerous the activities that communities carry out with external associations and groups of neighbors are, or even how some areas of the community are made available to the neighborhood. This tendency toward open relationships in today's territories is not obvious and should, in my opinion, be seen as one of the most relevant aspects of co-housing.

The second aspect concerns the ability to innovate. In fact, co-housing is a relatively recent and innovative phenomenon, not only because of the residential supply it offers, but also because of the way it relates to the environment. Therefore, it is in continuous evolution and able to constantly propose new solutions. This is demonstrated by the richness of the forms in which it can be presented, and it may be the fact of being raised as a proposal outside the traditional schemes that gives it this ability to adapt and to continuously regenerate itself. Perhaps these characteristics of innovation and taking on different forms are now generating a dilemma that is not irrelevant in regard to co-housing. If the first co-housings were born to offer human-made housing solutions, far from the optics of market and individualism, today it is the market itself that is increasingly interested in the world of co-housing and that does not conceal its objective of entering in it. This interest can be observed and evaluated in different ways. On the one hand, there are those who see it as a "field invasion" by a problematic force, such as the market, in a field on a human scale that wants to remain intact. On the other hand, there are those who consider this market trend to be a natural evolution of co-housing that, starting from "pure" experiences, comes to assume new conformations and is now contributing to social changes. Perhaps a characteristic of co-housing are the wish and desire to provide an example for society? An example to follow and imitate, capable of changing society? If we think about it, with this "field invasion," the dimensions of "sharing" and "environments to promote social relations" are, once again, offered on the market.

I believe, therefore, that a great dilemma arises as to whether co-housing must preserve its "original purity" or "contaminate itself" to have a greater impact on society. To this dilemma I think it is right that everyone gives their own answer.

References

1. Belterra Co-Housing. Welcome to the Belterra Cohousing website!. http://www.belterracohousing.ca/.
2. Bloomington Cohousing, Common spaces & amenities (2019). https://www.bloomingtoncohousing.org/common-spaces. Accessed 17 July 2019
3. Cannock Mill Cohousing (2019).. http://cannockmillcohousingcolchester.co.uk.
4. Cohousing.it, Cohousing Chiaravalle – Milano (2017). www.cohousing.it/portfolio/cohousing-chiaravalle. Accessed 14 May 2019

References

5. Coflats Badbrook, Home (2015). https://www.coflats.net. Accessed 22 June 2019
6. Archdaily, 1–6 Copper Lane N16 9NS/Henley Halebrown Rorrison Architects. Archdaily (2014). www.archdaily.com/580881/1-nil-6-copper-lane-n16-9nshenley-halebrown-rorrison-architects
7. L. Tebbutt, Henley Halebrown Rorrison uses timber and brick for London's first co-housing development. Dezeen (2014). https://www.dezeen.com/2014/09/21/copper-lane-co-operative-housing-henley-halebrown-rorrison-london/. Accessed 14 February 2019
8. Cohousing.it, Cosycoh – Milano (2017). http://www.cohousing.it/realizzazioni/cosycoh/. Accessed 15 June 2019
9. Cranberry Commons, Cranberry Commons Cohousing (2014). www.cranberrycommons.ca. Accessed 14 June 2019
10. Doyle Street Coousing, Doyle Street Cohousing, Emeryville, CA (2013). http://www.emeryville-cohousing.org. Accessed 23 May 2019
11. E. Giorgi, Man and environment: looking for the future. PhD thesis, University of Pavia, Italy (2016)
12. Earthsong Eco Neighbourhood, Welcome to Earthsong Eco-Neighbourhood (2014). https://www.earthsong.org.nz/home. Accessed 17 May 2019
13. Ecosol (2014). www.ecosol-fidenza.it. Molto più di un condominio: un luogo abitato. Accessed on 11 May 2019
14. Emerson Commons Cohousing, Home (2018). http://www.emersoncommons.org. Accessed 3 July 2019
15. Lancaster Cohousing, Welcome to Lancaster Cohousing (2019). http://www.lancastercohousing.org.uk. Accessed 7 June 2019
16. Frogsong, A Cohousing Community (2013). www.cotaticohousing.org. Accessed 15 May 2019
17. Cohousing il Mucchio, Sognando si realizza la realtà (2014). http://www.cohousingilmucchio.it. Accessed 24 June 2019
18. S. Sitton, Un'oasi in un deserto di nebbia. Irughegia (2011). https://irughegia.wordpress.com/2011/11/14/unoasi-in-un-deserto-di-nebbia-4/. Accessed 22 May 2019
19. K1, K1 at Marmalade Lane (2014). http://www.cambridge-k1.co.uk. Accessed 30 April 2019
20. LILAC, Low impact living affordable community (2019). http://www.lilac.coop. Accessed 6 June 2019
21. Los Portales. http://www.losportales.net. Accessed 12 May 2019
22. Milagro Cohousing, This is our Milagro (2017). http://milagrocohousing.org. Accessed 13 June 2019
23. Moora Moora. Co-operative Community, Diverse community (2018). http://mooramoora.org.au. Accessed 29 May 2019
24. The Mount Camphill Community – Independent Specialist College & Cohousing, Cohousing (2016). http://www.mountcamphill.org/default.aspx?PageID=9. Accessed 20 June 2019
25. Munksøgård, Welcome to the website of Munksøgårds (2013). http://www.munksoegaard.dk/index_en.html. Accessed 29 May 2019
26. Tam Associati, Cohousing: shared living in Bologna (2015). http://www.tamassociati.org/portfolio/cohousing/. Accessed 27 May 2019
27. Nubanusit Neighborhood & Farm, An old-fashioned neighborhood in a new way (2018). http://www.nhcohousing.com. Accessed 2 June 2019
28. Old Women Co-Housing, We've built our own community (2017). http://www.owch.org.uk. Accessed 24 May 2019
29. Pacific Gardens Cohousing Community. Welcome to Pacific Gardens Cohousing Community in Nanaimo, Vancouver Island, BC. http://www.pacificgardens.ca. Accessed 21 May 2019
30. Pioneer Valley Cohousing. Our Community. https://www.cohousing.com/welcome. Accessed 14 May 2019
31. Cohousing Pomali, Unser Zuhause (2015). http://pomali.at/unser_zuhause/index.html. Accessed 11 June 2019

32. Tam Associati, Cohousing: shared living in Treviso (2015). http://www.tamassociati.org/portfolio/cohousing-shared-living-experience-2/. Accessed 12 April 2019
33. Quayside Village Cohousing, Quayside Village (2011). https://sites.google.com/site/quaysidevillage/home. Accessed 23 April 2019
34. Radiance Cohousing (2019). https://www.radiancecohousing.ca. Accessed 20 April 2019
35. Rancho la Salud Village, Discover Rancho la Salud Village (2017). http://rancholasaludvillage.com. Accessed 21 April 2019
36. Rocky Hill Cohousing, Welcome to Rocky Hill Cohousing (2014). https://www.rockyhillcohousing.org. Accessed 12 March 2019
37. Solidaria Cohousing a Ferrara. Il progetto a Ferrara (2017). http://www.cohousingsolidaria.org/page.asp?menu1=3&menu2=11. Accessed 7 March 2019
38. Springhill Cohousing, Springhill Cohousing Welcome (2017). http://www.springhillcohousing.com. Accessed 8 March 2019
39. Stoplyckan. Hjärtligt välkommen till Kollektivhuset Stolplyckans hemsida! http://www.stolplyckan.nu. Accessed 12 March 2019
40. SCI du Tournesol, Our Community (2018). https://sunflowercohousing.org.uk/our-community.html. Accessed 8 June 2019
41. Swan's Market Cohousing, History (2019). https://www.swansway.com/getting-it-built/. Accessed 22 April 2019
42. Cohousing California, Temescal creek cohousing (2018). https://www.calcoho.org/temescalcreek. Accessed 23 May 2019
43. TerraCielo, Il Complesso (2016). http://www.terracielo.biz/complesso.php. Accessed 12 February 2019
44. Freiburg-vauban.de, Vauban (2016). https://freiburg-vauban.de/en/quartier-vauban-2/. Accessed 14 February 2019
45. Central Women, Welkom in de Wandelmeent (2017). http://www.wandelmeent.nl. Accessed 12 February 2019
46. Wind Song Cohousing Community, Homes & Environs (2019). https://windsong.bc.ca. Accessed 30 June 2019
47. Wolf Willow Cohousing. History of Wolf Willow Cohousing. https://wolfwillowcohousing.ca/our-project/. Accessed 26 March 2019

Chapter 6
Hopes

The previous chapters present a series of facts and considerations with the intention of promoting reflection on co-housing, on the reasons for its worldwide success and on its impact on contemporary society; these last pages want to collect personal considerations on a vision of hope in which the design and practice of co-housing can take on a decisive role in restoring the lost alliance between humanity and the environment—a hope on the role that design can (and should) assume in the management of social and environmental emergencies. The thesis that guided the drafting of this book is that the success of this operation passes through a renewed awareness that people are responsible for the environments in which they live and that this awareness is renewed starting from the regeneration of the community environments, whose design is a fundamental moment. Without doubt, this historic alliance today has been lost. Our time is marked by radical transformations and changes in all the forces that have characterized Western thought, culture, and society. Contemporary society is facing important challenges. First, the information technology revolution and globalization, which have already changed relations and the market, extending them beyond the borders of countries and embracing the entire Earth. Numerous crises developed, bringing with them obscure and abstract fears. Continuous economic alarms and environmental concerns arouse widespread individualism and the withering of human relationships. Cities increasingly feel the effect of a society that promotes only individualism and relationships with digital reality: urban territories became like archipelagos of islands divided by ethnicity or wealth, characterized by the inability to recognize the common needs and the common resources to share [1].

A fearful perception of being colonized and oppressed by hardly identifiable forces is spreading within a powerless society, which seems to have lost control of its decisions and the freedom of its actions. The certainties vanish—what dominates is science, which by substance is based on continuous negations of itself; moreover, the center of the world still exists, but now it is not connected to a specific territory [2]. We are experiencing a serious crisis that has hit humanity for a century now. It is a heavy collapse of both the humanities used to organize and explain the world,

© Springer Nature Switzerland AG 2020
E. Giorgi, *The Co-Housing Phenomenon*, The Urban Book Series,
https://doi.org/10.1007/978-3-030-37097-8_6

and the tools that man has always used to stabilize his actions in the world and to define his position in relation to the environment.

So, while cities were once places of protection from wild nature, today they are considered a threat of devastation of nature itself. Technology is prevailing by providing humanity with tools that can be decisive or highly destructive. In this situation, technology—and no longer Utopia—is the force that transforms the way in which humanity relates to the environment.

First, for Greek culture, nature was an unchanging background that always existed, from which people deduced the laws for the right government of the soul and where they were placed with technical abilities that, however, could do nothing against the Ananke of Nature. Meanwhile, in Jewish–Christian culture, nature was seen differently, a product of the divine will, assigned to the protection of humanity, which was placed at the top of it. So here, the technique was defined as a domain over nature, as a procedure that had remained constant for almost two millennia, unable to subvert the natural order. Thus, modern science looked at the world with an unprecedented look, technical, and questioning: with the origin of technoscience, in the seventeenth century, there was no longer contemplation of nature, but only its manipulation. The weight of human actions on the environment became enormously more suffocating and qualitative values gave way to quantitative ones; operation gave way to execution; the *homo faber* gave way to the creator of homo. Gods disappeared and nature has proved fragile.

Until the eighteenth century, the impact of the human activity on the environment had never produced greater disturbances than that caused by natural phenomena. The emergence of industrial machinery, which made it possible to exploit natural resources more efficiently and injected never-before-seen amounts of gas in the atmosphere, represents a radical change: the entry into the new geological era of the Anthropocene. "Anthropocene" is a relatively recent term, but—to have a minimal understanding of the contemporary age—it cannot be ignored: it highlights, like no other term, the profound crisis reached by the relationship between humanity and the environment, but focuses on the immense human capacity to transform the environment, without underlining the limited awareness of how people can manage this immense capacity.

Later, the term "Anthropocene" was therefore changed to "Technocene," to highlight two factors: (1) the immense power with which man is impacting on the environment, mainly due to the technological capacities available today; and (2) the unawareness of the technological power available, and consequently of how to use this immense power, leads to a very dangerous loss of that sense of responsibility toward the environment.

Therefore, with the term "Technocene" we wanted to encourage a reflection on these issues, in order to highlight a state of deep crisis characterized by two dangerous factors, unawareness and irresponsibility. In contrast, awareness and responsibility were the founding pillars of pre-industrial societies, because the technical possibilities were not the technological ones we have today, and, moreover, those societies had strong, accepted religious and ethical principles, which gave a sense of belonging to the individual. Quoting Heidegger today, the alarming aspect is not that the world is turning into a huge, unique technological apparatus, but that we are not prepared for this transformation and, even more disturbing, that we do not have an alternative thought to the technological one. With the uncontrolled development of technological systems and the contemporary weakening of religious and ethical principles, awareness and the sense of responsibility have weakened. More than any other has been lost the awareness of belonging to the environment surrounding us: thus pushed an individualistic system, like the one we are living in, makes a person's attention limited to a single cage of individuality, losing sight of the environmental context. It is for this reason that one of the most common hopes of the Western world is the necessity to find again the lost equilibrium with the environment, defined in an alliance capable of guaranteeing wellness and happiness for the people of today and for the generations of the future. But is it possible, nowadays, to create again a man-oriented environment and rediscover an environment-oriented humanity? Recovering the awareness of being an integral and fundamental part of the environment is the first initial step. People must rediscover that this environment is both natural and social and, therefore, rich in important common goods that require a responsible attitude of protection on the part of all.

It is in this context that the meanings of sharing and design take value. The walls of individualism break up when a person feels that they belong to a group and when they are sharing common goods with a community, and design has the potentiality to catalyze this process, even if, before, an intense reflection on contemporary design is necessary. Firstly, we must be conscious that, in the last century, design and architecture became victims of a rationalizing mania; nowadays architecture answers more to the requirements of a regulatory and technological world than to the natural function to create spaces where this alliance can be expressed, which cannot be fixed by strict rules. So it cannot be forgotten that architectural design must answer to the changing nature of man, proposing spaces for experimentation and invention [3]. Secondly, we cannot forget that design, in its noblest assumption, is a process of awareness and deep understanding of the reality, to the point of also being considered a "research product" [4]. In this perspective, citizens and their ideas must be placed at the core, as in the project Open Your Space, in Shanghai, which is based on the idea that "the value of places goes hand-to-hand with the emergence of a new idea of a sustainable well-being" [5]. For these reasons, designers must be able to give a strong signal of change: the "dumb buildings" must leave space again to the "singing architectures" [6]; the landscapes of intensive exploitations must leave space again to Francesco Petrarca's landscape; in the fighting between ideological man and technological man, design processes must help the ideological man to increase the work and to defend the dignity of people. This is the direction that design must retake if we really want to give a new hope. In a stratified context, we must build spaces for individuals and communities, where we can rewrite an updated alliance between humanity and environment. To do this, design must restart from history, rediscovering again the beauty of the individual, with their social values and their involvement in a community life, where sharing must be the aim and means for a new way of making architecture.

In this perspective, nothing is to be invented, but everything is to be rediscovered. History teaches us that the presence of a system of shared values and the enhancement of a healthy identity define the community, a place where people live in a relationship defined by ethical and natural reasons. If history teaches us how to create community, architectural practices must be the main protagonist of these more than ever necessary transformations, and the role of technology must not be demonized, since in the digital world we can find useful tools for improving design and civil participation. As observed by Ni Minqing, if technology is conceived as a public good, cities will be able to provide strategic and visionary proposals and be really "smart." We can, and we must, design open processes to "engage citizens in collective participation for digital social innovation" [7]. Some works as careful about human happiness as about social and environmental sustainability are already being realized, projects that grow from the bottom up, from people to people, based on sharing and attention to the environment. In this context, co-housing is spreading with great success, creating interesting and lively housing realities. Co-housing is a way of sharing life that combines the private sphere with common spaces and collective moments; it is a contemporary rediscovery of an old way of living that has always existed in the human living custom. In recent years, this phenomenon is spreading in the developed countries where people perceive a strong need to restore a people-friendly living style and sharing values, today lost due to the frenzy of consumerism. The designing interest for this topic is not generated just by the new meaning of the living spaces, but also by its capacity to involve the context, bringing innovations and vitality to difficult realities. Several research works propose the activity of sharing as a strategy for solving the problem of this unbalanced relationship between individuals and the environment [8]. The fact that sharing can help to recreate a new context, where the oppression created by the technological apparatus is controlled and the decisions can be led by human will and dream, is confirmed by several cases.

Recreating physical networks among people and reducing the level of material inputs needed to perceive happiness are both possible thanks to the enhancement of public and common goods. The feeling of belonging to a group is fundamental for the wellness of the individual because it is strictly part of our being humans: sharing moments, looking for social and ethical relations and the instinctive estrangement of the ghost of loneliness are, and always have been, the aim of the living solutions. The idea of "*civitas*," still visible in the historical cathedrals, is nowadays lost but, as architects, we have the duty to create environments in which this idea of public enhancement and communitarian belonging could be realizable. Architects and designers must be brave enough to rediscover it. We can consider for a final time the meaning that individual and social values assume in driving the life of individuals and communities, precisely because the loss of values jeopardizes that awareness and responsibility that we spoke of before. Today moral imperatives come from technology and they make the moral postulates of our ancestors seem ridiculous, not only those of individual ethics but also those of social ethics.

An alternative is urgent—the ethics that should drive human behavior is weakened so an enormous spectrum of considerations on the actual condition of soul, conscience, and morality can be opened. Can the soul, such a rich and thoughtful element, coexist with the technological apparatus? Can consciousness and morality drive human action once again? Alternatively, can human actions, by now contemporaries of technology, do anything other than follow the path marked by the self-reinforcing of the technological apparatus? We must escape from the increasingly widespread and alarming danger of the indifference and greyness of habits, which are the enemies of values and beauty. Just like Luciano Fontana, who cut his canvases to leave a glimpse of the art beyond the monotony of his paintings, so design and communities can cut the indifference and the habits of daily life to rediscover a new system of values that allows us to rewrite a renewed alliance with the environment. This is not an easy challenge, but if we do not lose the ability to dream and to hope, humanity can win and, in these times of change, design is one of the most powerful weapons in our hands.

References

1. A. Bugatti, *Discomfort of the City. Thoughts Under Construction* (Maggioli Editore, Santarcangelo di Romagna, 2017), p. 51
2. T. Cattaneo, *Study on Architecture and Urban Spatial Structure in China's Mega- Cities Suburbs* (Universitas Studiorum S.r.l, Mantua, 2016)
3. A. Bugatti, *Contaminatio. Between Art and Architecture* (Maggioli Editore, Santarcangelo di Romagna, 2018), p. 33
4. R. Amirante, *Il Progetto come prodotto di ricerca* (Lettera Ventidue, Siracusa, 2018)
5. M. Ni, Open your space: a design-driven initiative in Chinese urban community, in *Open Your Space. Design Intervention for Urban Resilience*, ed. By M. Ni, M. Zhu (Tongji University Press, Shanghai, 2017), pp. 82–179
6. P. Valéry, *Eupalino o l'architetto* (Mimesis, Sesto San Giovanni, 2011)
7. M. Ni, M. Zhu, *Open Your Space. Design Intervention for Urban Resilience* (Tongji University Press, Shanghai, 2017)
8. T. Cattaneo, R. De Lotto, *Rural-Urbanism-Architecture. Design Strategies for Small Towns Development* (Alinea Editrice, Florence, 2014)

Index

A

Africa
  Matmata
    adaptation, 74
    Arab and Middle East building
      tradition, 76
    building typology, 75
    geographical area, 74
    identity and community, 76
    patio, 76
    settlements, 75, 76
    strong cultural connotation, 77
    Tunisian subterranean communities, 75
    underground constructions, 74
  Umuzi/Kraal
    activities, 78
    aspects, 77
    historical settlement, 77
    incorporeal environment, 78
    inhabitants, 79
    Zulu society, 77
    Zulu villages, 78
Age of technology, 22–24
America
  Calpulli
    administrative center, 81
    chilampas, 80
    colonial *aciendas*, 81
    definition, 79
    settlements, 80
    social relations, 80
    societies/ethnic groups, 79
    Tenochtitlan, 80
    territorial management, 80
    territorial organization, 79

  Raramuri settlements
    agricultural production, 82
    aspects, 82
    Chihuahua, 81
    housing settlements, 83
    natural environment, 82
    North American natives, 81
    Sierra Tarahumara, 82, 83
    storage and exciding, 82
    teaching, 82
  Shabono
    Amazon forest, 84, 85
    belongings, 84
    consensus, 84
    distribution system, 84
    timber structure, 84
    Yanomami, 84
Anthropocene, 260
  definition and history, 15–17
  earth overshoot, 18
  responsibility, 18, 19
Arab and Middle East building tradition, 76
Architects and designers, 262
Architectural approach, 113
Architecture, 44
Asia
  Kibutz
    common facilities, 72
    "communal settlement", 71
    configurations, 73
    contemporary society, 73
    cooperative lifestyle, 71
    Jewish roots, 71
    kibbutzim, 73, 74
    "past" cases, 71

© Springer Nature Switzerland AG 2020
E. Giorgi, *The Co-Housing Phenomenon*, The Urban Book Series,
https://doi.org/10.1007/978-3-030-37097-8

265

Index

Asia (*cont.*)
  "perimeterial" position, 72
  total equality, 73
  Tulou
    aspects, 68
    Chinese buildings, 70
    Chinese empire, 70
    cultural relations, 71
    earth buildings, 69
    environmental equilibriums, 69
    horizontal distribution system, 70
    political power, 70
    residential buildings, 68
    threats, 70
Aztec Empire, 80, 81

B
Berber-Arab community, 75
Building methods, 37
"Bustling small city", 69

C
Calpultin (Nahua plural of *calpulli*), 80
Chihuahua, 81
Chilampas, 80
Chinese buildings, 70
Civitas, 10, 11
Co-housing, 259, 262
  characteristics, 107–109
  community, 112
  contemporary territories, 99–101
  context, 111–114
  description, 99
  design, 112, 113
  designer, 104
  final list, 118, 119
  first list, 117, 118
  four stages, 96–98
  functionalist perspective, 98
  gated communities, 106–107
  housing, 112, 113
  "industrial revolution", 95
  levels of sharing, 110, 111
  "living community", 96
  living together
    collaboration and networking, 94
    contemporary issues, 93
    contemporary society, 95
    economic-juridical principles, 94
    panorama, 94
    "traditional house", 95
  phenomenon, 114, 119
  projects (*see* Projects, co-housing)

real estate, 105–106
reasons, 110
resilience, 102–103
reuse, 101–102
second list, 118
six features, 109
sociological perspective, 98
traditional conception, 117
CoHousing Israel (CHI), 138
Collective buildings, 62
Colonial *aciendas*, 81
Common goods, 31, 59, 261
"Common" ventures, 59
Communitarian and environmentally friendly
  based approach, 247
Communitarian environment, 262
  common goods, 30–31
  community, 31–34
  humanity, 27
  industrial community, 34–36
  sharing, 28–30
Communitarian project, 109
Communitarian settlements
  members relationships, 88
  physical dimension, 88
Community, 31
  blood and soil, 33
  building, 86
  contractual, 53–55
  definition, 32
  industrial, 34–36
  movements, 29, 35
  sense, 33
  social capital, 31
  typologies, 29
Community Supported Agriculture
  (CSA), 197
Consciousness and morality, 263
Contemporary society, 259
Contractual communities, 93, 100, 106, 110
Cosmological interpretation, 89
Cultural approach, 42

D
Design approach, 40, 101
Dwellings (iQukwane), 79

E
Earth and social drying, 19
"Earthsong Centre Trust", 151
Eco-friendly approach, 121, 133,
  140, 155
Economic-juridical principles, 94

Index

Economy and environment, 19–20
Environmental context, 261
Environmental crisis, 2
Environmentally friendly approaches, 98
Environmental sustainability approaches, 209
Ethical approach, 40
Europe
  cascina
    agricultural work, 66
    farm phenomenon, 65
    farmhouse typology, 65
    public activities, 66
    rural settlements, 64
  Phalanstery
    architect Fourier, 66
    building codes, 66
    Falansterio, 67
    Fourier, 68
    gallery, 68
    Seristeri, 68
    settlement, 67
    vision, 66
    wage system, 66
  polis
    The agora, 64
    collective buildings, 62
    communitarian ideology, 62
    Greek *poleis*, 61
    Greek society, 61
    harmony, 63
    ideas, 62
    moral value, 62

F
Falansterio, 67
Five architectural ways
  commons, 45–46
  contractual communities
    classification by Brunetta and Moroni,
      53–54
    classification by Sapio,
      54–55
    territorial, 55, 56
  co-working, 50–52
  "dead" and "alive" architecture, 45
  icons, 48–50
  public space, 46–48
Francesco Petrarca landscape, 11, 12

G
Gated communities, 89
Greek culture, 260
Greek society, 61

H
Historical architecture
  communitarian experiences, 60
  community, 61
  dormitories/social housing, 61
  features, 60
  identity, 60
  projects, 59
  purpose, 59
Hi-tech production method, 51
Human activity, 260
"Human dilemma", 16
Human environment
  architecture and soul, 39–41
  field of housing, 36
  inhabitants live with soul, 41–42
  principle of responsibility, 42–44
  rationality, 37–39
Humankind
  crisis, 3–5
  environment, 1–3
  progress, 5–6

I
Ideological approach, 51, 106
Individualism, 259
Industrial approach, 35
Industrial Revolution, 60, 95
"Institutional experiments", 53
"Integral urbanism", 40

J
Jewish–Christian culture, 260

K
Kibbutzim, 73, 74
Kibbutz Movement, 73
Koinon-koinonia, 62

L
Laws of nature, 14
Lisbon earthquake, 14, 15

M
"Machine" approach, 41
Maoris
  architectural complex, 86
  communities, 86
  community building, 86
  complex relationships, 86

268 Index

Maoris (*cont.*)
  culture, 86
  genealogy (*whakapapa*), 88
  god of war, 87
  indigenous population, 85
  pre-Christian societies, 86
  taiao, 85
  welcome ceremonies, 87
  western vision, 87
Marae, *see* Maoris
Mathematical–geometric laws, 93
Meson, 62

N
National Geographic report, 70
North European welfare approach, 230

O
Operational method, 37

P
Philosophical approach, 41, 121
"Political community", 32
Principle of public interest, 43
Principle of responsibility, 42–44
Private Gated Communities (PGCs), 106
"Project for Public Spaces" (PPS), 47
Projects, co-housing
  Belterra, 120, 121
  Bloomington, 123–125
  Cannock Mill, 125, 127
  Casa Tucuna, 129–131
  CHI, 138
  Chiaravalle, 131, 133
  Coflats, 135, 138
  Copper Lane, 139, 140
  CosyCoh, 140–142
  Cranberry Commons, 144, 145
  Doyle Street, 145, 146, 148
  Drivhuset, 149, 150
  Earthsong Eco-Neighbourhood, 150, 151, 154
  Ecosol, 154, 155, 157
  Emerson Commons, 157–159
  final list, 120
  Forgebank, 159
  Frog Song, 162, 163
  Il Mucchio, 165, 166
  Itaca, 168, 169
  K1, 169, 171
  Le Torri, 171, 172, 174
  LILAC Grove, 174, 176
  Los Portales, 179, 181

  Milagro Co-Housing, 181–183
  Moora Moora, 183, 185, 188
  Mount Camphill, 188, 189
  Munksøgård, 190–192
  Mura San Carlo, 192, 194, 195
  Nubanusit, 195, 196
  OWCH, 198, 199
  Pacific Gardens Co-housing Community,
    199, 202, 203
  Pioneer Valley, 203, 204
  Pomali, 205, 207
  Quattropassi, 209, 211
  Quayside Village, 211, 212
  Radiance Cohousing, 213–215
  Rancho La Salud Village, 215, 218, 221
  Rocky Hill, 221
  Solidaria: San Giorgio, 224, 225
  Springhill, 225, 227, 228
  Stolplyckan, 228, 230
  Sunflower, 231, 232
  Swan's Market, 234, 235
  Temescal Creek, 236, 238, 241
  Terracielo, 241
  Urban Village Bovisa, 243, 244
  Vaubandistrict, 246, 247
  Wandelmeent, 247, 248
  Windsong, 249, 250
  Wolf Willow, 252, 253, 255
Public and common goods, 262
Public space, 46–48

R
Rationalist approach, 38
Renowned sustainable approach, 205
Research product, 261

S
Science and industry
  changing society, 12
  control and dominate nature, 14
  laws of nature, 14
  Lisbon earthquake, 14, 15
  man *vs.* environment, 12
  market, 13
  religious technology, 15
  states, 13
  time, 13, 14
Scientific method, 6, 13, 15
"Scientific revolution", 13
Seristeri, 68
Shared living
  Africa, 74–79
  America, 79–85

Asia, 68–74
Europe, 61–68
Oceania, 85–88
"Short Twentieth Century", 37
Social and environmental emergencies, 259
"Social capital", 31
Social relations approach, 155
Societies and environment
civitas, 10–12
Greece, 6–8
humankind, 6
Judeo-Christian culture, 8–10
Subjective well-being (SWB), 29
Sustainable approach, 113
Syssitia, 62
Systemic approach, 32

T
Tarahumara, 81
Technique and technological apparatus, 21–22

Technocene, 24, 260
Tenochtitlan, 80
Territorial community, 55, 56
"Trace gas", 17
Traditional approach, 41
Traditional communities, 89

W
Welcome ceremonies (*pōwhiri*), 87
Western civilization, 62
"World of cities", 10

Y
Yanomami, 84

Z
Zulu settlements, 77
Zulu society, 77